WRITING POWER

SUNY series, Studies in Scientific and Technical Communication
James P. Zappen, Editor

WRITING POWER

Communication in an Engineering Center

Dorothy A. Winsor

STATE UNIVERSITY OF NEW YORK PRESS

Published by
State University of New York Press, Albany

For information, address State University of New York Press,
90 State Street, Suite 700, Albany, NY 12207

Production by Diane Ganeles
Marketing by Jennifer Giovani

Library of Congress Cataloging in Publication Data

Winsor, Dorothy A.
    Writing power : communication in an engineering center / Dorothy A. Winsor.
    p. cm.
    Includes bibliographical references and index.
    ISBN 0-7914-5757-5 (alk. paper) — ISBN 0-7914-5758-3 (pbk. : alk. paper)
    1. Communication in engineering. 2. Technical writing. I. Title.

TA158.5.W56 2003
620'.001'4—dc21
                                                          2002030481

10 9 8 7 6 5 4 3 2 1

CONTENTS

ILLUSTRATIONS

Acknowledgments

I once heard a new graduate student puzzling over the emphasis that we in rhetoric place on collaboration, community, and the influence of texts on one another. He argued that, at this point in his education, he should know how to write. I found his argument understandable and even remembered thinking along the same lines. I find, however, that the longer I work as a scholar, the more I depend on my colleagues to help me and push me along. My work on this book was greatly helped by feedback from Ann Blakeslee of Eastern Michigan University, and from colleagues from the Group of Researchers in Applied English Studies at Iowa State University: Rebecca Burnett, Carol Chapelle, Dan Douglas, Barb Duffelmeyer, Carl Herndl, Lee Honeycutt, David Russell, and David Wallace. Two anonymous reviewers from State University of New York Press helped me to see further possibilities in the work I had drafted. And of course, I am more grateful than I can say to the people at Pacific Equipment Corporation for allowing me to become a temporary part of their work life.

An earlier version of chapter 4 appeared in *Written Communication* (17.2), pp. 155–284, © 2000 by Sage Publications. Reprinted by Permission of Sage Publications. Portions of chapter 3 appeared in the *Journal of Business and Technical Communication* (12.3), pp. 343–370, © 1998 by Sage Publications. Reprinted by Permission of Sage Publications. Portions of chapter 5 appeared in the *Journal of Business and Technical Communication* (15.1), pp. 5–28, © 2001 by Sage Publications. Reprinted by Permission of Sage Publications.

Vignette 1

Scenes in an Engineering Center (and Elsewhere)

Greg is an engineer in Pacific Equipment's engineering center. When I arrive to observe him today, his cubicle is, if anything, more buried in paperwork than it was the last time I was here. In response to a question from me, he says that his desk always looks like this. Engineering drawings are spread out on several surfaces. Posted on the walls are a copy of the periodic table, a chart showing the heat and oil resistance of elastomers, a hazardous warning chart, and a Forbes for President button.

He checks his voice mail, making notes in his planner as he listens. One call is from a coworker who wants Greg's advice on a part he is including in a vehicle he is designing. Greg calls him back and leads him through a series of questions: "What's it going between? Plastic? Urethane plastic? If it wasn't a quarter inch thick, if it was close to an eighth of an inch thick, would that be acceptable?" As he listens to the answers, he makes more notes. He tells his coworker to come to his cubicle to pick up a catalog that will show where he can get all of the parts he needs. He has to give his coworker directions on how to find his cubicle, even though they work on the same floor.

* * *

Jim is a technician in Pacific Equipment's labs. Today he is rebuilding a machine that is used to test vehicle water pumps. The work on this machine has a long and convoluted history that Jim explains to me as he works. At one point, an engineer wrote a work order

1

requesting that "backlash" be set to a certain distance. Backlash is the distance between the teeth of one gear and those of the gear with which it interlocks. It represents the distance a tooth would move before it meshed with an opposing tooth if the gear were reversed. The technician who was working on the machine at that point told a student-intern who was working with the engineer, that the backlash could not be set to the specified distance because of the way in which a part was machined.

Jim does not know what happened to communication after that, but apparently the engineer did not realize that the backlash had not been set. The pump was run for over 500 hours of testing. Jim took over the testing midway through the procedure and ran the pump for several days before he noticed and then told the engineer that the gear train sounded noisy. The engineer wrote a new work order for Jim to measure and set the backlash to what he had originally requested. At that point, Jim discovered the impossibility of setting the backlash as requested and told the engineer. The engineer told Jim to measure the real backlash, and then, to salvage his 500+ hours of data, he was going to compensate mathematically to figure out what he really had versus what he thought he had. Jim has taken his measurements and is now reassembling the pump to run a further 1,500 hours of tests.

* * *

Ken is a manager in the engineering center. He is holding his weekly meeting with the team leaders who work under him. Fourteen men and I sit around the table. As the first order of business, Ken introduces me. They joke about what I might be learning about how not to write at work.

Most of the team leaders have notebooks and one has a laptop. Everyone takes some sort of notes during the meeting. Ken runs the meeting using a small notebook in which he has listed topics to be covered. The purpose of this meeting appears to be less discussion by participants than information dispersal by Ken. He begins by reporting on a safety program that has apparently led to a lowered accident rate in various company facilities. I have seen posters referring to this program throughout the engineering center. (I have also attended a meeting where, as part of the safety program, we all watched a bloody, absolutely terrifying video on the dangers of fork trucks.) Ken emphasizes that everyone must take the safety class, though he realizes that they are all pressed for time. He asks everyone to agree to complete the class; one person asks for a schedule. Someone else says that the schedule is in an E-mail from the previous Friday, and the person asking for the information confesses that he hasn't waded through all of his E-mail yet.

Next, Ken talks about the pricing of a Pacific Equipment vehicle, which is apparently more expensive than a competitor's version. Ken is not asking these team leaders to do anything about the cost of this vehicle, but rather to keep the cost issue in mind as they supervise the work of engineers on their teams. As the meeting progresses, I note much talk about competitors, who apparently serve as a kind of bench mark against which Pacific Equipment measures itself. Money and budget concerns also come up repeatedly.

* * *

Mark is an engineering student who is working as a summer intern at Pacific Equipment. His project is the remodeling of two test cells. We meet in the lobby at 6:55 A.M. and go immediately to a meeting room in the laboratory area where, Mark says, he has a project meeting every morning. At the meeting are Mark, his boss, another Pacific Equipment engineer, and three employees from the contractor doing the remodeling. Everyone (including me) has earplugs, safety glasses, safety shoes, and hard hats. Mark repeatedly consults a notebook in which he has spreadsheets that show the status of the many tasks associated with the remodeling.

They discuss their progress on remodeling projects. The contractors tested some of the equipment yesterday. Mark asks if they tested the automatic shutdown, and one of the contractors says, "Yes, first thing." They all laugh. This is a joke because it means that the automatic shutdown engaged immediately, which is a sign of a problem with the equipment. In contrast to the meeting of Ken and his team leaders, this one seems to me to be rather egalitarian. The participants all contribute and seem to listen to one another. The meeting ends about 7:30. Two of the contractors talk to me as we leave, asking me why I'm there. One says that E-mail is important and that I can't understand what's going on unless I have access to it.

* * *

Alan is an engineer designing a new vehicle. He has been working with another engineer named Bob to modify an engineering drawing. I have watched them as they take turns drawing on the same sheet of paper. They have decided to use a handbrake being developed for another Pacific Equipment vehicle, but they don't know the part number. Alan is searching for the number in the manuals in his cubicle, but he is having trouble finding it. He tells me that the handbrake's number is not in the company's on-line catalog. He will check this catalog for another division, he says, but that division doesn't post part numbers if the part is at an early design stage.

Alan suggests to Bob that they look at a prototype vehicle in the company's labs, but Bob says that the technicians are probably testing it today. Alan phones the labs and asks, "Are you running in the sound room today? OK; we won't disturb you." He tells Bob they can look at the prototype during the technicians' 9:00 break, but he's afraid they won't see much anyway because the area in which they are interested is covered. He asks Bob if he has seen what the technicians did to make an experimental part workable. He makes a sketch of the technicians' solution to show Bob. The technicians apparently used some cut and weld improvisation that amuses the engineers at the same time that they admire its ingenuity.

* * *

I sit at the desk in my home office and read the fieldnotes I have created to record these scenes. I wonder what they mean about creating communal knowledge within organizational power structures. Because I am a rhetorician, I wonder about the role of texts such as engineering drawings, notes in a planner, posters, spreadsheets, E-mail, work orders, manuals, catalogs, and budget sheets. What role do texts play in the power that groups of people are able to exercise over (through?) their own knowledge-making activity and that of others upon whom they rely?

CHAPTER 1

USING WRITING TO NEGOTIATE KNOWLEDGE AND POWER

For the last several years, I have wondered about the relationship between knowledge and writing, particularly as it occurs in the work of engineers (see Winsor 1990a, 1990b, 1992, 1994). What is the relationship between the ability to create knowledge and the ability to represent that knowledge in various symbol systems? For instance, how do engineers use verbal, mathematical, graphic, and other symbol systems in their day-to-day work? What is the relationship between the ability to embody knowledge in an object, to represent knowledge in texts, and to generate knowledge in a group? In one sense, the answer to this question is obvious. Engineers use texts to think with both individually and, much more often, in groups (Henderson 1999). They share disciplinary texts such as data curves and test reports so that they can jointly determine the meaning of those texts and thus understand the object they are designing and testing. In vignette 1, we see the engineers Alan and Bob jointly focusing on an engineering drawing in order to achieve exactly this kind of common understanding. Presumably, in this use of texts, engineers are typical of most people who work together in groups. Texts function not only to record and share what is already known but, perhaps more importantly, to help writers and readers generate and agree on what is to count as knowledge.

Indeed, engineering is a prime example of what we have come to call "distributed cognition." Edwin Hutchins (1993, 1996) provides the most commonly cited example of distributed cognition in his study of crew members navigating a ship. (See also Freedman

5

and Smart 1997.) As Hutchins makes clear, distributed cognition does not mean just that different people accomplish different parts of a task, but that people and their tools acting in concert can interact to accomplish a kind of cognition that no individual could achieve separately, and indeed that such interaction is probably necessary for some cognition to occur at all. For instance, people acting jointly can provide a more complicated kind of reasoning than any one person could maintain. Thus, a group of engineers can maintain a variety of theories that contribute to their design work, whereas an individual would have trouble seeing from a similar multiplicity of perspectives at once. Distributed cognition can also make cognition more robust, so that if one person is unable to accomplish a task (e.g., to interpret a data curve), in some cases someone else can step in and complete it. So in vignette 1, Greg is able to help a coworker select a part that will work well in a design, making that design better than either engineer could have accomplished on his own. Thus Hutchins argued, "cognitive accomplishments can be joint accomplishments, not attributable to any individual" (1993, 35).

Moreover, in watching a group of people navigate a ship, Hutchins came to recognize "the extent to which the computational accomplishments of navigation are mediated by a variety of tools and representational technologies" (1993, 35). Tools such as maps are not incidental to navigation, but rather are essential to it. We think in conjunction with other people and with our mediating tools, and if we are engaged in complex, distributed cognition, such tools are indispensable because they mediate between things and people (the map mediates between the navigation team and the port), between various groups of people (the map mediates between the harbormaster's crew and the crew of the ship), between individuals (the map mediates between the crew member plotting the ship's position and the crew member on deck taking a positional reading), and within the activity of individuals (the map guides the crew member's understanding of his or her own activity). That is, texts, in this case maps, can be crucial mediating tools for thinking even in an individual. We see such a mediating text in vignette 1 when Mark, the summer intern, uses a set of spreadsheets to guide his understanding of the progress being made on remodeling the Pacific Equipment labs. In addition, for people who are working

Texts vs. Symbols

jointly, texts can serve the additional important function of communicating and encouraging people to think together. Indeed, Hutchins concluded that communication was a crucial factor in allowing a system of distributed cognition to function. So quite clearly, one aspect of the relationship between text and knowledge is that the former is a tool for the production of the latter.

However, such an analysis is incomplete because it ignores the circumstances in which much knowledge work is done, that is, in for-profit, hierarchical corporations. Certainly, most engineering knowledge is generated within such systems of power and profit. While Hutchins's (1993, 1996) examination of distributed cognition calls our attention to the fact that knowledge is often communally held and thus depends on communication, we must remember that systems of distributed cognition are not always collaborative, egalitarian, and harmonious. Hutchins's own shipboard example is obviously a hierarchical one. Thus in addition to asking how texts and knowledge interact, we also need to ask how both of those factors interact with power. How does an organizational hierarchy affect how people at various levels can cooperate to create knowledge? Within a hierarchy, people in more powerful positions are often able to determine what knowledge is valuable and even what facts or ideas are to count as knowledge for the organization. So vignette 1 shows us the manager Ken making it clear that knowledge about competitors and cost is to be valued and acted upon. We need to ask how people in a hierarchy create and use texts such as reports, drawings, budgets, and E-mail to build knowledge together and to have it validated by those in positions of power within the organization.

hierarchy

Moreover, if, as Michel Foucault (1980) claims, power is not a quality that anyone can "hold" but a relationship that is always locally generated using means that include, but are not limited to, positions in a hierarchy, then how are the generation of knowledge and text connected to the generation and enactment of power? Hutchins's work implies that in systems of distributed cognition, social arrangements are also knowledge work arrangements. The creation of knowledge is enabled to proceed in some ways and constrained from proceeding in others by the way in which people interact. As part of those social arrangements, the power arrangements that flourish in any organization affect how both text and

social
economic

knowledge are produced and used. Conversely, we should expect that knowledge and text are among the resources that can be deployed to create relations of power. So, for example, we see in vignette 1 that, at Pacific Equipment, the routine use of work orders makes it far easier for an engineer to tell the technician Jim what the engineer wants than for Jim to tell the engineer about problems with the procedures he or she requested.

In other words, knowledge, text, and social structures are intertwined. Over the last fifteen years, rhetoricians have increasingly treated the intertwining of text and social structures in terms of genre. Carolyn R. Miller (1984) was the first to define genre as a form of social action, a typified textual response to a typified social situation, and not simply a collection of repeated formal characteristics. Her work has been developed by such scholars as John Swales (1990), Aviva Freedman and Peter Medway (1994), and Carol Berkenkotter and Thomas N. Huckin (1995). Their work and that of others (e.g., Pare 1993; Schryer 1993; Smart 1993; Blakeslee 2000) has provided us with an increasing number of studies describing how genres function in various social situations. However, rhetoricians have only begun to explore the way in which power is a factor in the creation and dispersal of genres. JoAnne Yates's (1989) historical study of railroad companies deploying various forms to control their employees is an early example of such work. Yates demonstrated that "regular flows of upward, downward, and lateral communication as well as detailed record-keeping procedures played a critical role" in establishing the systematic management toward which companies aspired at the turn of the nineteenth century (1989, 2). More recently, Bernadette Longo (2000) has also conducted a historical examination of the way in which scientists used technical writing to generate its position as our era's dominant knowledge. She argues that the fact that technical writing is designed to divert attention from itself to the subject matter is one of the factors giving it power: "The invisibility of technical writing attests to its efficiency as a control mechanism because it works to shape our actions without displaying its methods for ready analysis" (2000, ix). In many studies of writing in the workplace, questions of power are implicit (and here I would refer to most of my own work). However, increasingly they are explicitly examined, as in Anthony Pare's (2000) study of the way in which social workers' reports must remain subservient to those of physicians in hospitals, Susan M. Katz's (1998) examination of how an organizational newcomer was able to generate a powerful position for herself, or Carl G. Herndl's (1996) discussion of a biologist's resistance to the dominant discourse in his workplace.

Freedman and Medway exemplify this interest in genre's relation to power when they argue that we need to ask questions such as these:

> How do some genres come to be valorized? In whose interest is such valorization? What kinds of social organization are put in place or kept in place by such valorization? Who is excluded? What representations of the world are entailed? The absence of such questions is the ideological limitation we see as most needing to be addressed in the next stage of genre studies. (1994, 11)

In this book, I respond to Freedman and Medway's call for a more ideologically aware examination of genre by arguing that part of the social action implicit in using a genre is related to power. By this I mean that the typified nature of genres encourages certain actions and discourages others. For example, when companies require periodic activity reports from engineers, the latter make every effort to have activities to report. Similarly, when performance review forms include space where supervisors can comment on the actions of subordinates but not vice versa, the expectations built into the genre discourage looking to inept management as a reason for poor performance by a subordinate. And as we see in vignette 1, when the written genres that Pacific Equipment associates with testing are limited to the work order the engineer writes to start the tests and the test report the technician returns when the testing is complete, there is no routine way for the technician to send written notice about problems in a test's design. Thus technicians must rely on oral communication and information is more easily lost as a result.

As structures, genres are one of the resources that Anthony Giddens refers to when he talks about how people exercise power within social systems such as organizations. Resources, he says, "are structured properties of social systems, drawn upon and reproduced by knowledgeable agents in the course of interaction. . . . Resources

limitations

Genre

Miller's definition

are media through which power is exercised, as a routine element of the instantiation of conduct in social reproduction" (1984, 15–16). As social actions, and as typified responses to typified social situations, genres are a "routine element of the instantiation of conduct." In organizations, they usually preexist the user, although any user can vary them and thus contribute to their slow change. As part of the historical being of any organization, genres are a resource that can be deployed, manipulated, contested, and regulated. They are thus a way in which power is constructed, organized, and put into effect. In this book, I will suggest that organizations tend to institutionalize genres that reinforce existing power relationships so that not all of the writing that people do is equally likely to be recognized as part of an organizational genre. I will also claim that in their institutionalized form, organizational genres are not equally available to everyone in an organization. Rather, they become resources only for those who are authorized to use them.

In the chapters that follow, I talk about how power, generic texts, and knowledge interact in an engineering organization, hoping that understanding a single case will aid us in creating a picture of how they may interact in other settings. While I believe that engineering organizations share many characteristics with other kinds of workplaces, I also believe that they are particularly interesting sites in which to examine the connections of knowledge, power, and text, because they straddle the boundary between science, which is usually identified as knowledge work, and commerce, where relationships of power are taken for granted. Thus they are an ideal setting for examining the questions I am raising:

- How do texts, and particularly generic texts, help people in various organizational roles to generate knowledge?
- How are texts used to create and occupy positions of power? What does this use of texts tell us about genre?
- How does power affect the generation of text?

In examining these questions, I take a somewhat different tack from much of the literature on organizational communication. This literature often seems to fall into one of two camps. On one hand, much literature in business communication seems to assume that

implicit

organizational communication is primarily a unidirectional, top-down dispersal of information and instructions that shape the actions of subordinates. This literature often takes the organizational chart as a representation of power relationships that are not open to question. Many of the articles in the *Journal of Business Communication*, for instance, seem to operate from this assumption. For example, recent issues include articles on how companies can most effectively introduce quality programs so that subordinates will accept them (Lewis 2000) and how organizations can communicate strategically to maintain their legitimacy in times of crisis (Massey 2001). Research on such topics is useful if organizations are to function well, but if our research is limited only to such topics, we miss much of the reality of how texts function in organizations. The study I describe in this book departs from this work in that it examines organizational communication happening at many levels, in all directions, with a variety of purposes that sometimes conflict and may all be seen as valid. In other words, I want to argue that the generation of power through discourse should be examined, that its existence (and rightness) should not be assumed, and that texts play a role in the way in which power is created and deployed.

In contrast to this first strain of work that seems to assume the legitimacy of hierarchical action and to ignore communication initiated by subordinates, a second strain of work grows out of critical theory and treats capitalist organizations as exploitative and power as automatically oppressive. In rhetoric, this strain of work shows up most often in journals like *College English*. For instance, recent issues have included articles assuming that "resistance" to current social structures is automatically desirable (Muckelbauer 2000; Wehner 2001). I believe that capitalism's primary valuation of profit often does lead to inhumane behavior on the part of organizations and that power can be misused. But I also believe that for-profit organizations often do useful work and that power is a way to accomplish that work. To quote Giddens, "structure is not to be equated with constraint but is always both constraining and enabling" (1984, 25), and "power is the means of getting things done, very definitely enablement as well as constraint" (175). Thus power has a dual nature, a fact that much of the research on organizational communication ignores. In this book, I assume that power always exists and that understanding the textual operations of power in organizations

like
Sponsors

will be useful for everyone involved. This study aims at clarifying how power and knowledge are textually created and controlled in the material reality of the social system within which we live.

GENRES IN ACTION IN AN ENGINEERING CENTER

The specific site for this study was the engineering center of Pacific Equipment, a large manufacturer of off-highway equipment.[1] Pacific Equipment's engineering center contains the development facilities for two different divisions: Off-highway Equipment Engineering, which designs and develops the vehicles that the company produces, and Engine Engineering, which designs and develops the engines that are installed in various company vehicles. An imaginary tour of the engineering center will give us a preliminary glimpse of some of the genres in action there.

The Test Labs

When we enter the engineering center on our imaginary tour, we first encounter the cafeteria that is conveniently located in the center of the building. After we sign in at the security desk, we turn to our right and enter the labs that extend in a sprawling, single-story area that constitutes two thirds of the facility's floor space. In this space, lab technicians build and test prototype vehicles, engines, and components. For instance, the labs contain dynamometers, a sort of treadmill for engines that can be used to test a new or improved engine's durability or to measure its emissions. They also contain a huge cold room, where vehicles can be left for several days at below zero temperatures and then (the designing engineer hopes) started, and an equally large sound room, where the technician can measure how loud a vehicle is. As we walk through this area, we see many signs painted on walls, doors, and floors. They admonish us, for instance, to wear hearing protection or to walk only inside certain lines so that we avoid being hit by the small carts and fork trucks that occasionally cruise down the hallway. However, the genres that are the focus of the most attention by people in this area are those that circulate between it and the engineering area that is on the other side of the cafeteria. These are the work orders that come from the engineers, laying out the technicians' tasks, and the test reports that technicians return to the engineers, giving the results of various procedures. As we walk through the labs, we see technicians consulting work orders and entering test results into their computers for shipment back to the engineers.

This book will be looking at three technicians who performed a variety of tasks in the lab:

- Gary,[2] who ran tests on new Pacific Equipment products,
- Jim, who constructed experimental parts, and
- Rich, who ran a supply center from which other technicians could get parts and built custom-designed test equipment.

Gary, Jim, and Rich all used work orders. Gary and occasionally Jim also submitted test reports to the engineering area.

The Engineers' Cubicles

On our tour, we follow these electronically submitted reports back across the cafeteria and into the engineering area. The engineers' cubicles are arranged in a three-story structure, with the top two floors devoted to vehicle engineering and the bottom floor to engine engineering. In this area, engineers create designs and analyze data from the labs. The cubicles for members of any engineering group are usually arranged in the same area making it relatively easy for engineers to overhear and observe one another's work. Additionally, in the areas of most engineering groups, a table is also available for informal group meetings.

The typical cubicle overflows with paper. Texts are filed, posted, and stacked on desks. As we walk along an aisle of cubicles, we see engineers studying their computer screens, which display texts in a variety of genres that have been E-mailed to them. Most crucially for the generation of engineering knowledge, we see them studying the data from lab tests that technicians have shipped to this side of the engineering center or the drawings of parts they are designing and will eventually instruct the lab technicians to build and test. While some engineers are doing this individually, many work jointly with colleagues to create or interpret these engineering genres. They crowd around a single computer screen or sit across from one another at the same desk and jointly arrive at some sense of what a drawing should look like or of what data tell them.

Thus creating and interpreting texts serve as ways for engineers to negotiate joint knowledge with one another that they will then attempt to persuade their colleagues or supervisors to accept. They prefer to communicate their conclusions orally in meetings, but to their dismay, must periodically lay aside the ongoing work in which they are currently interested to prepare written reports for their managers. Drafts of these reports, too, are displayed on some of the engineers' computer screens. While engineers usually don't like to write these reports, they know they need to take care with them if they are to persuade managers to allow them to do the work they would like to do.

This book will focus on five engineers:

- Dan, who designed and analyzed the results of tests of drivetrain problems;
- Greg, who served as a consultant advising other Pacific Equipment engineers who might want to use "compliant" material (i.e., material such as rubber that can be deformed without losing its properties) in parts they were designing;
- Dave, who analyzed the structures that design engineers had created and verified whether they were strong enough;
- Alan, and
- John, who were both engineers designing new vehicles.

Additionally, like most engineering organizations, Pacific Equipment commonly employed college students to work for them in the summer, intending, if all went well, to hire the students once they graduated. On our tour, we occasionally see one of these summer students laboring over drawings or data and consulting with the more experienced engineers who mentor them.

Team Leaders' Cubicles

The supervisors for these engineering groups, whom Pacific Equipment calls "team leaders," are in cubicles adjacent to those of their groups. In their cubicles, we see them reading reports from various engineers in their group, drawing together the information, and preparing presentation visuals that they will use as part of their technical progress reports they must submit to their managers. In these presentations, they will also ask for resources and justify the use of the resources they already have. If they are not in their cubicles, they are probably in one of the meeting rooms ranged in a row down the center of each floor of the engineering area. We will see two team leaders in the pages of this book:

- Brad, who was a team leader for sensor development although when I observed him, there were no other members of this group, which was only six months old; and
- Paul, who was a team leader for electronics and led a group of approximately six people.

Managers' Offices

Near the meeting rooms where team leaders spend much of their time, we also see the offices of upper-level managers. Both the managers' offices and the meeting rooms have doors and walls that extend to the ceiling and thus offer more privacy than the cubicles do. Managers are sometimes in their offices, reading a wide variety of texts that almost all arrive via E-mail. For upper-level managers, most of these texts have to do with allocating resources of various kinds. These resources include personnel (Whom shall we hire? How should we evaluate this employee? Into which group should we place this engineer?) and the time and money represented in budgets. Like team leaders, these managers are frequently absent from their offices attending meetings where they decide how work at the engineering center will progress.

We will see two upper-level managers in this book:

- Ken, who was a director at the engineering center and performed a similar function for a Pacific Equipment development center in another state; and
- Doug, who was a manager of Technical and Engineering Services, an area with almost three hundred employees.

When we look at all of the people in all of the areas, we see a system of distributed cognition held together partly by generic texts. As I already noted, Miller (1984) has argued that genres are social actions; they are typified rhetorical responses to typified social situations. In Pacific Equipment's engineering center, genres such as work orders, test reports, reports to managers, and budgets are used to carry out the center's mission to create engineering knowledge. They are simultaneously used to regulate the way in which various people interact to carry out that mission. The typification we see in most of these texts is actually mandated by the organization. Miller argues that people's perception that a situation is recurring leads to the formation of a genre. But when a genre has been institutionalized, as most of those at Pacific Equipment have been, then the recurring form of the genre can also be used to encourage people to perceive situations as similar and to behave in ways that the genre calls for. That is, genres can invoke a situation as well as result from it. Genres are always re-created in the way in which each person uses them, but in order to be a genre, they also have to hold onto structure tightly enough for colleagues to mutually recognize them. At Pacific Equipment, that structural repetition is often not accidental, but rather represents the belief of people in positions of relative power that some situation is and should be recurring. To some extent, genres are deployed to enforce that perception of repetition. Thus the genres that flow through the various areas of the engineering center represent a confluence of the creation of knowledge and the enactment of power.

KINDS OF CAPITAL IN ENGINEERING

This book, then, presents a case study focused on how one engineering organization uses texts to create and maintain its knowledge structure and the related question of how it uses texts to create and maintain its power structure. I use the word *structure* here, but it is important to remember that neither knowledge nor power is stable; rather they are both dynamic and shifting. There is no structure unless people constantly engage in structuring (Giddens 1984); there is no order unless they engage in ordering (Law 1994). Social actors are not simply puppets responding to the forces around them. The technicians, engineers, and managers I describe are not helpless pawns. Rather, they act, albeit with incomplete freedom, to shape the structures within which they then function. As this book will demonstrate, texts are one of the means by which people generate and stabilize both knowledge and power.

Because engineering knowledge and organizational power seem to reflect two different authority systems, data generated in the lab and hierarchy established by the corporation, we might expect that they would conflict. It would theoretically be possible, for instance, for a manager to ignore the recommendations that engineers generate from data and mandate that a cheaper but less sound product be built. However, in this book, I will argue that, at least in the highly successful Pacific Equipment Corporation, managers did not customarily use their authority in this potentially problematic way. Rather, power and knowledge tended to be converted into one another; according to Pierre Bourdieu (1991) different kinds of capital can be converted into one another. According to Bourdieu, capital or credit exists in different forms that can be expressed in terms of economic logic, although they are not reducible to money. In addition to monetary capital, for instance, he speaks about "social capital," which refers to prestige (e.g., to hierarchical positions within an organization),[3] and "cultural capital," which refers to cultural knowledge or competency (e.g., engineering knowledge). Under the right circumstances, Bourdieu says, these kinds of capital can be converted into one another. That is, for example, managers could use their authority to enable the generation of engineering knowledge and then use that knowledge to solidify their own positions as valued employees.

Subordinates, too, could trade power for knowledge and vice versa. For instance, at Pacific Equipment, the technicians responded to work orders that had to originate in the engineering area. No matter how good an idea the technician had for how work should be completed, he or she was not authorized to issue a work order. Such a situation is consistent with Bourdieu's (1991) assertion that some words will be effective only if the speaker or writer has been institutionally authorized to deliver them. However, engineers did sometimes consult technicians about how testing should be done so that the latter could affect work orders even though they were not authorized to write them. In vignette 1, the engineers Alan and Bob are admiring an improvised solution that technicians have created to a problem in building an experimental part. Engineers could sometimes use their engineering knowledge in reports that persuaded managers to allocate funds to carry out research in which they were interested, even though the engineers were not institutionally authorized to decide on the use of resources.[4] Thus knowledge and power were converted into one another. In this book, I will demonstrate that texts were often one of the means by which knowledge and power were converted at Pacific Equipment.

CONVERTING CAPITAL BETWEEN FIELDS IN THE ENGINEERING CENTER

In order to introduce the notion of using texts to convert different kinds of capital at Pacific Equipment, I want to draw upon Bourdieu's (1993) notion of "field." According to Bourdieu, a field is an arena of structured positions whose interrelationships are determined by the distribution of various kinds of capital. The distribution of capital can be formal (as when one person in an engineering area is designated a team leader), or informal (as when all members of an engineering group agree that one member is exceptionally knowledgeable). Bourdieu demonstrates that fields can be nested inside one another. For instance, the cultural field of art and literature exists inside the field of power (by which Bourdieu means economic or political power), except that the cultural field reverses the signs of success that exist in the larger field. That is, the value of literature and art in the cultural field is generally inversely related to their commercial success because highly sophisticated or cutting edge work is likely to have a small audience. A similar dynamic seems to exist at Pacific Equipment, in that the field of engineering exists inside the field of the for-profit organization but places value on well-designed objects even when they would be unprofitable to produce.

The engineer's interest in what my participants generally referred to as "quality" is both a resource and a problem for a corporation struggling to gain economic capital. It is a resource because a quality product will usually sell better. It is a problem because it needs to be made answerable to corporate concerns about cost and schedule. The problematic nature of the engineer's devotion to quality is reflected in the half-joking engineers' saying that it is sometimes necessary to Shoot the Engineer if a company wants to get a product out the door. Engineers are notorious for making costly improvements to products that managers have decided are good enough in their current incarnation. If we define a "successful corporation" as one that generates significant sums of economic capital, then such a corporation needs to manage the engineering value of quality so that it contributes to, rather than obstructs the accumulation of, economic capital, the means to power within the corporate world.[5]

How, then, do managers place engineering knowledge in the service of economic capital at Pacific Equipment? That is, how do they convert these kinds of capital into one another? In order to answer this, I want to draw on Bruno Latour's discussion of centers of knowledge and power (1987, 232–33). Latour says that Western Europe became a center of power because it had the technology to send explorers around the world and to bring back knowledge that could be amassed in one place. The technology to do this included not only ships and navigational tools, but also technologies of representation, such as writing and map drawing. By these technologies of representation, explorers were able to create "inscriptions" (64) whose force came from their ability to serve as "immutable mobiles" (227). That is, they were mobile because they could be moved from place to place and thus be amassed, but they were also immutable, so that when a map was moved from the South Pacific to England, it did not change. Knowledge had been temporarily stabilized so that it could be used. Power drawn from technologies of representation led to stabilization and accumulation of knowledge that in turn built power in centralized locations. Knowledge was converted into power that could then be reconverted into forming more knowledge.

Similarly, at Pacific Equipment, engineering knowledge is placed at the service of monetary capital by means of representations, that is, by means of texts. Engineers submit reports and various other documents to managers, a task that most of them find burdensome because it seems to be irrelevant to the work they are doing within the engineering field. Managers, however, operate in the corporate field and, from their point of view, these reports and similar texts allow them to decide what knowledge is to be generated and how knowledge is to be used, depending on whether a course of action will be profitable for the company. Requiring and controlling representations becomes a way to generate power for managers. And moving in the reverse direction, managers can deploy monetary capital to enable the generation of engineering knowledge if they believe that such an investment will return a profit to the company. This deployment of capital is regulated by means of another text, the budget sheet to which engineers must match their expenditures, so that managers do not lose control of capital even after it has been converted. Thus, we see an exemplification of Bourdieu's (1991) claim that, under the right circumstances, which here includes the presence of texts, various kinds of "capital" can be converted into one another, although the conversion will tend to be made to benefit the interests of those in positions of power. Monetary capital can be converted into cultural capital like knowledge or social capital like prestige that is intended in turn to yield greater monetary capital. Power is not used to trump engineering knowledge and to ignore its implications. Instead, power and knowledge become the means to create one another with texts as the mediating tools.

Indeed, Michel Foucault (1980) argues that power and knowledge are two ways of looking at the same thing, that they are, in fact, the same thing, which he calls "power/knowledge." Half of this insight is echoed in the cliché that knowledge is power, but Foucault also argues that power must exist if we are to recognize something as knowledge. Brenton Faber articulates Foucault's position in his analysis of discursive factors leading to organizational change. For Foucault, he says, "power comes before truth and before knowledge. We 'know' something only because we consent to the authority presenting the information. Accordingly, things are not objectively true or false, and knowledge cannot exist apart from relations of power" (2002, 114).

Fig. 1.1. *Institutionalized Genres and Power/Knowledge*

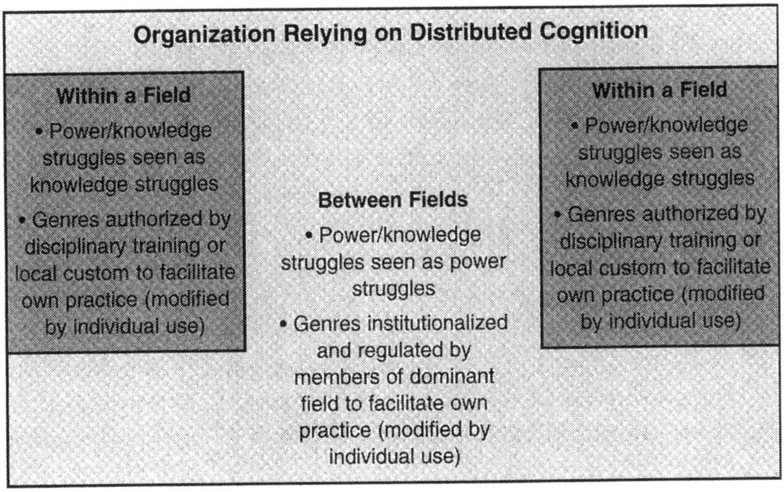

Organization Relying on Distributed Cognition		
Within a Field • Power/knowledge struggles seen as knowledge struggles • Genres authorized by disciplinary training or local custom to facilitate own practice (modified by individual use)	**Between Fields** • Power/knowledge struggles seen as power struggles • Genres institutionalized and regulated by members of dominant field to facilitate own practice (modified by individual use)	**Within a Field** • Power/knowledge struggles seen as knowledge struggles • Genres authorized by disciplinary training or local custom to facilitate own practice (modified by individual use)

INSTITUTIONALIZED GENRES
AND POWER/KNOWLEDGE

In this book, then, I will draw upon my observations of Pacific Equipment's engineering center to explore the relationship between genres, knowledge, and power illustrated in figure 1.1. The theory I illustrate grows from the fact that, within a large organization, people from various fields need one another in order to provide the distributed cognition that makes organizations more productive than individuals. Engineers need lab technicians and they also need managers despite the fact that neither of these groups shares the engineers' knowledge and disciplinary value systems. Within a given field, people tend to share disciplinary training and work experience. They therefore hold many assumptions in common. The generic texts they work with inside their field tend to be those they have learned in school or created with like-minded coworkers. So, for instance, engineers find it relatively easy to negotiate about texts such as an engineering drawing or a data curve. They don't necessarily agree about how the drawing should look or what the data mean, but they do agree on what counts as a valid argument and what evidence should be valued. In other words, they see the power/knowledge struggles internal to their field as struggles over knowledge.

In contrast, when people from different fields interact, they often operate from different assumptions about the importance of knowledge or action. They thus find it more difficult to agree on what standards should be applied to settle any disagreements by means of negotiation. In other words, in a situation that is the opposite of what occurs within a field, they see the power/knowledge struggles between their fields as struggles over power. In this situation, the genres in use can't very well grow out of disciplinary training or common experience. Rather, they are often those that the more dominant field has institutionalized and now requires. In using these institutionalized genres, however, people from other fields can try to use various rhetorical tactics to maintain control over their own work, a goal that everyone at the Pacific Equipment engineering center seemed to have in common.

mainstream

ORGANIZATION OF THE BOOK

In the chapters that follow, I will examine the interconnected generation of knowledge, power, and text in various parts of the Pacific Equipment engineering center. In each part of the organization, I will look at the genres in use and the way in which those genres work to generate knowledge and power. This study is unusual in that I was able to look at writing done by people at many different levels and examine how the writing worked to connect them to others in the organization. I begin by looking at managers, the study participants who held organizational positions conventionally believed to be the most powerful, and move to groups that would be successively lower on an organizational chart. As the careful wording of the previous sentence suggests, being "most powerful" or "lower" are conditions that are complicated. People can generate different kinds of power at all levels of an organization, although managers' organizational positioning can provide some of them with resources for such generation that are not universally available. Moreover, I had some trouble deciding on the relative ranking of technicians (whom I discuss in chapter 4) and summer engineering interns (whom I discuss in chapter 5). The technicians often knew more than the interns did about the working of Pacific Equipment vehicles, but the interns were positioned as engineers even when they were not yet ready to do engineering. On the other hand, the technicians were fully fledged employees while the interns were not. I discuss the technicians before the interns primarily because doing so makes more obvious some of the ambiguities of the interns' positioning.

The discussion will also move from more abstract genres such as project schedules and budgets to those such as work orders that are more tied to the details of daily work and action. It is no accident that managers used the more abstract genres because their task was to determine the general direction that the work of the engineering center should take. Subordinates were responsible for supplying managers with information that helped them to decide on this direction and for transforming these general plans into specific actions that they carried out or delegated and then reported on. Thus their texts tended to be more concerned with designing and interpreting the work of individual people and devices. The further one moved through the engineering area and out into the lab, the more specific the texts tended to be.

Chapter 2 will look at the way in which managers used texts to shape and regulate the work of engineers so that various kinds of capital—symbolic, cultural, and monetary—were kept in balance. Its primary claim will be that people value and seek to enact genres that allow them to control the kind of capital that matters most to them. Chapter 3 will examine how engineers saw the processes of gathering data from the laboratory technicians, of generating knowledge among themselves, and of validating it with their managers. It thus shows people selecting different genres depending partly on the relation between their own organizational positions and those of their readers. It also shows them trying to shape genres that managers have institutionalized to serve their own purposes. In other words, it considers power relationships as part of the generic situation. Chapter 4 will show how a particular text—the work orders that engineers wrote for lab technicians—was used to draw on the knowledge of the technicians while reasserting the corporate hierarchy that tended to treat the technicians as tools. As part of showing how hierarchy was preserved, it will also show that the work orders' representation of the technicians' labor misrecognized the way in which technicians did their work. Chapter 5 will examine how summer interns gained access to knowledge by becoming members of the social structure of the organization. In other words, it shows how social structures and knowledge structures overlap and how newcomers' assumption of roles in these structures inherently involves questions about power. This chapter also takes up the question of the role of tools in shaping genre and other aspects of organizational life. In chapter 6, the concluding chapter, I will then summarize the way in which the material in these chapters relates to the questions I am asking.

Between the chapters are vignettes of life in the engineering center that are edited versions of my field notes. They focus on the parts of the engineering center that will be examined in the chapters following them, and, in those chapters, I draw on the preceding vignette to illustrate my points. I hope that these more narrative accounts of life at Pacific Equipment will provide a different kind of understanding than the more analytic chapters and will provide further examples of material that can be examined. In vignette 2, we see a manager using texts to work through issues having to do with personnel and budget and thus directing the efforts of subordinates toward goals that have been determined.

Two Hours in an Afternoon
of a Manager: Doug

It's 1:00 on the Monday afternoon after a long July 4 weekend when I arrive at the engineering center. When I enter the building, I'm in the cafeteria with the security desk to my right. The security guard greets me by name, has me sign in, and issues me a visitor's pass that I must wear around my neck. Then he calls Doug's office so that someone can come and get me and escort me to Doug's office. I'm not allowed to wander around the engineering center without an escort, both because there is potentially dangerous machinery that could create liability for Pacific Equipment if I were injured and because the center generates proprietary knowledge that is sometimes sheltered from visitors.

Doug comes and gets me and leads me back to his office, which is built of the same cubicle modules as everyone else's, but his has a door and walls to the ceiling. He has a circular table in one corner with four chairs. The desk is L-shaped with overhead bookshelves. On the desk are both a desktop and a laptop computer. The desk contains stacks of paper, a phone, and photos of Doug's family. Posted are one sheet reading "Insanity: expecting change while doing things the same old way" and another with his job description, which reads:

My job:
Communicate "our" vision/direction
Work with T[echnical] and E[ngineering] S[ervices]
    Leadership to determine what is to be done and do
    our best

Clarify expectations and boundaries
Produce the fullest possible integration with the product
    design groups
Support and enable the division
Help secure resources
Provide information and information sources
Help T and ES leaders attain personal satisfaction.

Doug begins his afternoon by reading his E-mail. He gets about forty E-mail messages per day, but he tells me he worked yesterday afternoon (i.e., Sunday of the holiday weekend) and so is caught up. The first message he opens is a performance review that he wrote at the request of an employee. Employees have to get evaluations from several sources, including one who would count as a "customer," or user of the employee's work. Doug counts as an internal customer for this employee, to whom he will eventually send the evaluation. He has filled out the required form but has had trouble sending it via E-mail. He has written to the employee who requested the evaluation, asking him for help on sending the form and has done a trial send, but he will wait to send it until he's had a chance to look at it again. "It's too important," he says.

After Doug has read a few more messages, he and I go to see one of his team leaders about budget decisions. The team leader has been at Pacific Equipment for six months. Doug says that he needs a detailed list of budget expenditures by project and month, for the rest of the fiscal year, which ends October 31. The team leader says that it's hard to do the figures on a monthly basis. Doug coaxes him along, giving him a spreadsheet that he had prepared. The team leader studies the spreadsheet and then takes out a file, saying that for the project that he "really wants to keep," a certain amount of money has been set aside. Doug asks if that reserved money shows up on the spreadsheet. The team leader says that money is listed but that the dollar total differs from what he has recorded in his file. Doug explains that the discrepancy exists because part of the project is being funded from a different source, something the team leader didn't know. Doug asks if the money can be spent by the end of the year. There's been a budget cut and the engineering center wants first to cut its budget by the amount that would not have been spent anyway.

The team leader says that he also wants funding for a customer survey. He has been waiting to do the survey until Pacific Equipment's customers will have completed some seasonally driven activity and vehicle owners will have time to answer questions. He and Doug discuss whether the funds that have been appropriated can be spent. Doug says that the first time the survey is run it will count as an overhead project being done to test the survey process. Subsequent surveys would be paid for from a different budget.

The team leader then talks about a different project he wants to keep and several others that have been "vaporized," he says. "They're not gonna happen." He and Doug consult the spreadsheet to see a zero next to these items. Doug then tells the team leader that budgeting is already starting for next year. The team leader somewhat indignantly says that he got an E-mail today telling him that his capital budget is needed by Wednesday. That's "poor management," he says. They talk a little bit about whether they need a certain type of high-tech microscope. Will they have employees who can use it? Could they contract someone else to do the work instead? The device is very expensive. But the team leader says that he has gotten at least 30 positive reactions from his engineers via E-mails about this device.

They gossip a little about where another group is getting money to do a project. As Doug and I get ready to leave, the team leader tells me that interpersonal communication of the type they have just engaged in is important but that E-mail is even more important because they do so much more of it. It's especially useful, he says, when people travel or when you have to communicate with someone whose schedule does not fit with yours. Doug interjects that he believes face-to-face communication is sometimes vital. The team leader admits that face-to-face communication is important at the beginning of a project but says that participants can then use E-mail just as well. Doug responds that he has seen E-mail that simply should not have been sent, presumably because it was intemperate. The team leader says that he misses the "recall" feature of the E-mail that was available at his old job. Doug says that he once saw two people engage in a lengthy argument over E-mail until he finally suggested that they meet face-to-face. As we leave, Doug asks the team leader to send him his budget requests via E-mail.[1]

Doug and I go up to the next floor of the engineering center to take up the same issues with a different team leader, one who has much more seniority at Pacific Equipment. At his cubicle, Doug asks if the team leader got his E-mail. The team leader says he got about fifty today and has not yet waded through them all. He says that people must have loaded E-mails onto their laptops over the weekend and then come in and sent them all this morning. He makes notes of what Doug wants. Doug gives him his spreadsheet and the team leader asks what format Doug wants him to use for the budget. They talk about what would be best. The team leader says he'll send the information Doug is asking for. Then he says that he can't access Doug's schedule on-line and they talk for a while about what the problem could be.

When we leave this team leader, we head for the laboratory. We stop outside the doors to the lab area so that I can put on appropriate safety gear from the cupboard that stands there. I put on boots that fit over my shoes and that have metal guards over the main part of the foot. I also put safety glasses on over my own glasses and put earplugs on a cord around my neck, so that they can be inserted when we enter areas requiring them. Because I have been here before, I knew enough to wear trousers so that my legs are covered and to leave any rings at home. Doug is already wearing safety shoes and glasses and has earplugs in his pocket. We go to the lab so that Doug can greet three people for whom today is a hiring anniversary. They have been working at Pacific Equipment Corporation for thirty-three years each. He tells me that he goes to the lab regularly because he wants to understand what technicians do and, he hopes, to boost morale. He also tells me, though, that on at least one occasion, a technician told him that he didn't have to bother congratulating him on his anniversary because Doug didn't know him from a hole in the wall anyway. Doug seems to accept such mild hostility as normal, although it is not the rule. On our way through the halls, someone stops Doug to talk about cars that have been set aside for employees to use when they travel between different Pacific Equipment facilities in town. Apparently, some people have been using them to get around on the engineering center grounds and they are therefore not available for their proper use. Doug tells the person to send him suggestions for new rules.

When Doug has delivered his last anniversary message, we finally leave the lab by the door next to the security desk, where I shed my protective gear so that I can leave for home. It's 3:00.

CHAPTER 2

MANAGING THE ORGANIZATION THROUGH POWERFUL TEXTS

A hot-air balloonist was blown off course. As he drifted, he spotted a man in the field below. He maneuvered his balloon until he was able to talk to the man. "Where am I?" he asked.

"You're about thirty feet over my field in a hot-air balloon," the man replied.

"You must be an engineer," said the balloonist.

"That's right," said the man. "How did you know?"

"It was easy," said the balloonist. "Everything you've told me is factually accurate and completely useless."

"You must be a project manager," said the man in the field.

"That's right," said the balloonist. "How did you know?"

"Because you don't know where you are or where you're going and although you've only been here for five minutes, already everything is my fault."
—Joke told by Pacific Equipment engineer

The development of institutions enables different kinds of capital to be accumulated and differentially appropriated, while dispensing with the need for individuals to pursue strategies aimed directly at the domination of others: violence is, so to speak, built into the institution itself.
—John B. Thompson, Introduction to
Language and Symbolic Power

In an organization like Pacific Equipment, as in most organizations in our culture, power is likely to be exercised in the form of symbolic exchange rather than in physical coercion. It thus becomes *symbolic power*, a term that Thompson says Bourdieu uses to refer not so much to a specific type of power, but rather to an aspect of most forms of power as they are routinely deployed in social life. For in the routine flow of day-to-day life, power is seldom exercised as overt physical force: instead, it is transmuted into a symbolic form, and thereby endowed with a kind of *legitimacy* that it would not otherwise have. (1991, 23)

As we might expect, then, one important means to symbolic power is language (see Fairclough 1989 for an extended discussion of the relationship between language and power). This chapter will examine some spoken and particularly written texts that Pacific Equipment managers used to generate and legitimate symbolic power in the engineering center. In particular, I will talk about budgets and documentation as institutionalized genres that allow managers to scrutinize and regulate the actions of employees without needing to engage in confrontation or open coercion.

This chapter will argue that people use or reshape existing genres to allow them to control the kind of capital that is most important to them. I do not mean important to them personally (although overlap is possible) but rather important to them in the positions they occupy within the organization. In the case of the Pacific Equipment managers, that capital is monetary, and a large part of their task is to control the generation of engineering knowledge so that it contributes to monetary capital. In their efforts to control important capital, those who occupy powerful positions within a group, such as Pacific Equipment managers, will often make the genres that are useful to them official so that they can be easily invoked. These official genres can then be pointed to as representing a form of social action that individuals try to match with their own behavior. Thus they function as a means to elicit cooperation from employees in constructing systems of knowledge and power.

When rhetoricians think about how power operates, one of the issues we confront is how to account for the simultaneous existence of agency and structure (Winsor 1999). We believe that individuals are agents who control their own actions; but we also acknowledge that people exist within social structures that predate them and shape what actions are possible. In terms of using texts to generate power, such a dual perspective means that agents can create, shape,

imbalance

and distribute texts as part of how they build a network to strengthen their own positions (Latour 1987). However, within an organization such as Pacific Equipment, some genres already exist before an employee comes to work there. Thus employees must work within this historically shaped textual environment when they write texts of their own. Moreover, employees are positioned with access to different genres and thus to different means to power. In any organizational position, an employee must use the means available with skill, or power will never materialize; but also in any position, some means are available and others are not.

I will begin this chapter by talking about the managerial task of getting all employees to focus on the same organizational goal, a joint focus that is necessary if symbolic power is to function. I will then move on to talk about budgets and documentation, two of the textual means managers used to accomplish this task. At Pacific Equipment, budgets were almost completely under the control of those in more powerful positions, even when other workers filled in the actual numbers that appeared in these texts, as Doug's team leaders are being asked to do in vignette 2. Thus budgets operated as a managerial genre, and because they distributed resources, they served to regulate the activity in which engineers engaged even in the absence of managers.

In contrast to budgets, documentation could be used by anyone in the organization in an attempt to shape the actions of others (Winsor 1999). However, only people in positions of relatively greater power could require that others complete documentation or that it be done in a certain format. Managers could institutionalize the form and frequency of documentation, and thus the work it represented. Thus managers, but not their subordinates, could insist that others write reports about their actions but choose to leave their own actions undocumented, a situation I will also talk about in this chapter.

power

GENERATING ACCEPTANCE OF
ORGANIZATIONAL GOALS

According to Thompson, symbolic power can operate only upon "a foundation of shared belief" (1991, 23). That is, both the more and less powerful people in any exercise of symbolic power have to share a belief that the joint activity in which they are engaged is worth doing and that the social organization of that activity is reasonable. At Pacific Equipment, one of the managers' functions is to create this shared belief in the common enterprise and in its reasonable order. For instance, in vignette 2, Doug's openness to having a subordinate suggest a policy on company cars is probably one small incident meant to help create the sense of a just, joint enterprise.

One important means by which managers create belief in a shared enterprise is through the use of texts. Managers use texts to create what cultural-historical theories of activity teach us to call a "common object" (Cole and Engstrom 1993; see also Wenger's 1998, 178–181, discussion of "alignment"). That is they must get everyone working toward the same goal with a minimum of conflict (Miettinen 1998). As shown in vignette 2, this task is named in a list that Doug, an upper-level manager, had posted in his office. The list was headed "My Job" and the first item on it was "Communicate 'our' vision/direction." As the quotation marks around "our" suggest, getting everyone to share the same "vision" can be difficult in organizations that are likely to consist of individuals and groups with overlapping but not identical interests. Putting this potential for difference in terms of discourse, Carl G. Herndl says, "if we only consider the authorized discourse, we oversimplify the complexity of cultural relations and overlook the diverse cultural and discursive practices that contest the authorized discourse" (1996, 456). Such ruptures in the authorized facade of discourse occasionally occurred at Pacific Equipment, as I will show when I talk about the fictitious elements in organizational charts and budgets and again in chapter 4 when I talk about the relationship of Pacific Equipment technicians to authorized texts. However, this company seemed to be reasonably successful in creating acceptance of organizational goals, a fact that may partly account for its success. Managers spent much of their time making sure that people worked toward a common goal despite conflicts between different interests, and they used texts as one means to achieve this commonality.

For Pacific Equipment as a whole, the object was, of course, corporate profit, achieved by producing and selling a desirable product. However, the immediate attention of the engineering center was focused on the design of that "desirable product." Within that term *desirable*, there were at least three elements:

- the product needed to contain features that the customer would want (an element that usually was referred to as "quality"),
- it needed to be producible at a reasonable cost, and
- it needed to be ready for the market quickly enough so that the company could compete with other manufacturers.

In the engineering center, quality was seen as the responsibility of engineers, while cost and schedule were perceived primarily as the responsibility of managers. These two sets of responsibilities were sometimes in conflict. As one engineer, John, said,

> There's always the conflict of interest. The engineer is often very focused on providing the features and the performance, and the manager looks at it often from a little different perspective and asks, "What is it going to cost? What's the reliability? What's the development and design time associated with getting this idea into production? How much does it delay my program?"

This is not to say that managers didn't care about quality and that engineers were blithely unaware of cost and schedule. Indeed, I will show that managers designed work practices to make sure that quality was upheld, and I saw engineers working on lowering the cost of a design they had created. Moreover, the performance appraisal form for engineers includes a category for "cost consciousness" and another for "meeting established and agreed upon deadlines" (a good example of an institutionalized genre being used to emphasize the goals of a dominant field). However, as John also told me, while engineers were concerned about cost and schedule, "they're not going to be held responsible in front of the general manager to say why their project is over cost." Managers and engineers are positioned differently within the engineering center and from their positions they are asked to account for different kinds of capital.

As can be imagined, designing a quality product at a minimum cost in a timely fashion was no easy task. Two different engineers jokingly told me that their managers could have any two out of the three but not all at once, which suggests that the tension between these requirements was commonly perceived. These requirements for quality and profitability occur at the juncture of the management and engineering fields and thus some tension is natural. While the tension between these requirements is intensely practical, it also exemplifies Pierre Bourdieu's (1991) point about how capital can be converted; Pacific Equipment can use its accumulated monetary capital to produce a product that is low cost, early in the marketplace, and/or high quality. All of these factors would shape the social capital that is built into the company's reputation. Monetary capital gets converted to symbolic capital (and vice versa). But capital is not endless, even for a big corporation. Trade-offs always have to be made, and the job of the manager is to control the form those trade-offs take. In other words, one of managers' primary jobs is to budget the various kinds of capital.

BUDGETING VARIOUS KINDS OF CAPITAL

Managers at Pacific Equipment began their task of balancing capital by organizing the work in the engineering center so that it was easy for engineers to enact the goals of management. The organization of work was, in effect, rhetorically done, because it acted to persuade employees to aim for certain common objects. In that sense, the organization is analogous to a text through which symbolic power is being generated.

Although Pacific Equipment is a hierarchical organization in which managers theoretically have the right to isue orders, by and large, managers there did not try to coordinate activities through orders. As John said: "It doesn't do any good to just go to your team of engineers and say, 'Take $500 out of this. I don't care how you do it. See you tomorrow.' I don't think that works well. You don't get the team energy. You don't get people's hearts behind it." Rather, as Bruno Latour would say, managers attempt to "translate" the interests of engineers so that they work toward the organization's interests in order to achieve their own (1987, 108–121). In obvious terms, as John pointed out, managers arrange the organization's reward system so that if the company does well, "we get a bigger bonus at the end of the year." However, engineers' interests were not reducible to "a bigger bonus" and management could not rely on money as the sole means to keep engineers' goals aligned with those of the organization.

Thus managers resorted to other means to structure engineers' work. Specifically, managers organized work activity so that employees were likely to enact management concerns in their own work even when managers were absent. In other words, among the most important texts for management were budgets and schedules, the texts that regulated cost and time. Managing was thus somewhat concealed. Indeed, one of the problems I faced when conducting this research was in identifying who, exactly, counted as a manager. When I asked engineers about their relations with managers, they frequently asked me whom I meant. The upper-level managers Doug and Ken counted as managers by anyone's standards, but what about the team leaders Brad and Paul? One engineer, Alan, belonged to a group that got a new manager, Jean, while I was observing him. Jean created an organization chart that also showed "team leaders" within the group. Should these team leaders count as managers? Alan didn't think so because, he said, they were more like "mentors," meaning their primary responsibility was to help the engineers develop the product. Thus they shared the engineering focus on quality rather than the management focus on schedule and cost, a factor that seemed to function as the dividing line between the fields of engineering and management.

However, this dividing line was not absolute because managers were able to push their concerns over cost and timeliness into the sphere of the engineers by the way they allocated resources. If managers put more money into a project, more time was devoted to it and it usually got completed sooner. On the other hand, if they invested less money into a project, engineers had to shape their search for quality to take very careful account of cost and available work hours, even when no manager directly told them to. Managers thus avoided the need to confront engineers directly. As Thompson says in the quote at this chapter's opening,

> The development of institutions enables different kinds of
> capital to be accumulated and differentially appropriated,
> while dispensing with the need for individuals to pursue

strategies aimed directly at the domination of others: violence is, so to speak, built into the institution itself. (1991, 24)

What this meant was that it was occasionally difficult to tell what counted as a moment when "management" was occurring. Management was not an outcome of people with organizational authority confronting and directing their subordinates. Rather it was constructed in the course of the normal activity of people engaged in regular work.

For instance, when I asked one engineer, Dave, if he had experienced situations in which managers and engineers disagreed, he readily answered yes, but when I asked him to give me an example, he described a conflict that had occurred between his engineering group and another group. Dave did not design vehicle parts himself but rather performed strength analyses in a department called "Product Validation and Verification." When an engineer in one of the design groups had designed a part, he brought it to Dave, who tested it to make sure it would be strong enough. Pacific Equipment deliberately allocated design and testing to different people because the design engineer has a natural tendency to believe that the part he or she has designed is a sound one.[1] Dave described a situation in which a design group refused to accept his judgment that a particular product design was not strong enough. This particular product was a completely new one for Pacific Equipment. Thus engineers lacked the knowledge that they had for existing products that were being improved. As it turned out, later testing proved that Dave's group was right and the product was redesigned.

When I pointed out to Dave that this example represented conflict between engineers rather than between engineers and managers, he disagreed and talked about the conflicting needs for quality, cost, and schedule. He saw the incident he had described as a conflict between quality and cost, and he attributed the concern for cost to the design group. He believed that he was less subject to time and cost "pressures" than the design engineers were and that having the parts brought to a service group was "how we address all three." Thus his identification of this as a conflict between engineers and managers depended on his perception of the managerial concern for cost having been built into the work of the design engineers. Man-

perception

agement was built into the way in which work was organized or, as Dave said, it was "built into the culture." If, as Clifford Geertz says, a culture is a kind of text, then the text of the engineering center is being used to generate relations of power and knowledge that remain in the control of managers. As Geertz also claims, seeing culture as a text is useful because it encourages us to ask how the text's (and culture's) meaning is achieved and solidified:

> The great virtue of the extension of the notion of text beyond things written on paper or carved in stone is that it trains attention on precisely this phenomenon: on how the inscription of action is brought about, what its vehicles are and how they work, and on what the fixation of meaning from the flow of events—history from what happened, thought from thinking, culture from behavior—implies. . . . ([1983] 2000, 31)

The text of the Pacific Equipment engineering center is arranged so that those within it will treat money and time as valuable kinds of capital that are appropriately handled by management.

We should notice that managers are not interested only in monetary capital, although that is the capital for which they are most responsible. Dave's analysis also suggests that managers want quality products. Thus they create a group like Dave's to frustrate their own desires for speedy, inexpensive design. They are aware that quality is important if the symbolic capital they get from their product's reputation is to be maintained. And a good product reputation is necessary if the flow of money into the organization is to continue. That is, various kinds of capital are being converted into one another and managers need to balance a "budget" in which all of them are maintained.

USING BUDGETS AND KNOWLEDGE
TO CREATE ONE ANOTHER

In the long run, then, managers are responsible for maintaining the budget of various capitals. But more immediately, their responsibility for cost means that they have to create and maintain budgets for the work at the engineering center. These budgets are powerful texts in the hands of managers. I had several opportunities to observe managers working with budgets, particularly when I was observing Doug, an upper-level manager. As vignette 2 shows, I shadowed Doug during a period when the following year's budget was being created. I also observed him at a meeting where a group of managers decided how certain items were to be categorized in their budget. These observations suggest the extent to which a budget is a text that converts the cultural capital of knowledge into monetary capital so that it can be regulated according to predetermined goals. Both engineers and managers were concerned about data but they cared most about different kinds of data. As the joke at this chapter's beginning suggests, data that seemed precise and relevant to the engineers could be considered useless to the managers. The engineers' work generated data about off-highway products, but to be useful for the managers, it had to be converted into data about costs.

Doug was the manager of both Technical Services and Engineering Services. Technical Services provided information like product-use data, simulations, analytic work, materials engineering, design analysis, and applied mechanics for the various product groups. Dave, whose work in Product Validation I just described, worked in this area. The other part of Doug's responsibility, Engineering Services, provided support for facilities that did not belong to any one group. This would include the test labs, the library, and office remodeling. His area was quite large, employing about seventy salaried people and two hundred hourly workers. As is common with managers, his task was to draw people together, coordinate their work, and support the work of others. As vignette 2 shows, while I observed him, Doug worked mostly with personnel issues (a sign of his responsibility for getting people to work together in the same direction) and budget. The emphasis on budget may vary somewhat by time of year but, for any upper-level manager, budget is always a concern.

A budget obviously represents power since a budget allows management to decide what is going to get funded. But at Pacific Equipment, a budget also represents knowledge in that the engineering center's actions are all aimed at generating knowledge. The monetary capital organized by the budget is meant to be converted into the cultural capital of knowledge. For instance, while I was observing Doug, budget cuts had been announced. In response, he was meeting individually with each of the team leaders reporting to him and asking them to cut their budgets and then to prioritize the rest of their projects. That is, the budget cuts quite clearly represented losses of some engineering projects. Some of these meetings are shown in vignette 2.

Doug's one-on-one meetings with his team leaders about the budget cuts are a good example of how a manager's power and an engineer's knowledge act reciprocally to affect one another. At first glance, budgets look like a top-down decision-making tool. But as I watched Doug and his coworkers and subordinates develop budgets, it was evident that, in making the decisions, managers took account of engineers' knowledge, a practice that was expected in the engineering center. Indeed, the second item in Doug's posted job description, right after "Communicate 'our' vision," was "Work with T[echnical] and E[ngineering] S[ervices] Leadership to determine what is to be done." In these one-on-one meetings, team leaders supplied their budgets, into which they coded their activities, to Doug, who, along with other managers, decided whether their proposals should be accepted or modified. Managers made these decisions knowing less about specific projects than the team leaders or their subordinates did, but they were able to do so because engineers' knowledge about the demands of the projects had been converted into the monetary language of the budget. Managers combined this coded form of knowledge with their own knowledge about wider "visions" (as Doug's job description calls it) in order to determine what work was to be done. Thus budgets were created through joint effort.

This is not to say that budgets were created in an egalitarian act. The conversion of engineering knowledge into managerial knowledge was obviously organized hierarchically, with engineers contributing to the decisions and managers implementing these decisions. As Edwin Hutchins points out,

> Gathering and providing information for the support of a decision are low-status jobs. Integrating information and making decisions are high-status jobs. There are, of course, exceptions. . . . In general, however, the goals are

in the hands of the higher status individuals—those who control the goals are, by cultural definition, of higher status. (1996, 204)

A budget then is a document that represents the cultural capital of engineering knowledge in terms of monetary capital, the capital that is of most interest to the corporation as a whole.

My observations also suggest that because budgets represent the point at which knowledge is converted into the language of monetary capital, creating them was not as straightforward and descriptive as I had assumed it would be. Coding actions into a budget was a classificatory rather than a descriptive process that helped to decide what kind of actions they were going to be understood to be. This classification was the provenance of managers operating within the confines of corporate and legal policies. For instance, the classificatory aspect of budgeting appears in vignette 2 when Doug explains that the customer survey will be budgeted differently the first time it is run than it will at all subsequent times. The survey is classified differently because managers believe that once the survey is established, it will generate knowledge about customer attitudes, whereas the first running of the survey will generate knowledge primarily about running the survey.

Classification was also evident at a managers' meeting on budgeting that I attended while shadowing Doug. The meeting had been called by the team leader in charge of *facilities*, a term loosely used to mean the building and everything attached to it. The point of this budgeting meeting was to decide whether certain items were to be counted and budgeted as "facilities," an act of categorization that did not seem to be straightforward. The team leader who called the meeting felt that some items were charged to his budget that should not have been. These included some materials for the test labs and work done by contractors to modify them so that tests requested by various design groups could be conducted. He wanted these costs to come out of the budgets of the product design groups rather than out of his because those groups decided what testing was to be done and thus spent money for which he was now held responsible. In other words, he was concerned about a situation that undermined the balance of different kinds of capital, leaving the generation of engineering knowledge unchecked by the need to respond to a monetary budget.

When attendees arrived at this meeting, they found a copy of the facilities budget laid out at each place around the table. This budget was a spreadsheet showing dollar amounts budgeted and those actually spent on facilities, month by month for each test cell currently in use. The team leader who called the meeting led a discussion that was governed by the text he had provided. He went through budget items he wanted to discuss, constantly pointing to items that were recorded on the spreadsheet. He had even highlighted all the copies in yellow to indicate the items he was concerned about. The text of the budget shaped the discussion, and seeing items recorded in the budget seemed to make those items available for scrutiny and discussion. "As you can see," said the team leader, "I'm spending less on [X] than I had budgeted," and "Go to the next page please, and you'll see. . . ." The meeting attendees could "see" what was happening through the lens of the text as focused by the team leader's comments.

The attendees at this meeting decided to try to reclassify the test expenditures and thus budget differently, although Doug told two team leaders at the meeting to consult with others to get agreement on how much to budget (a sign that budgeting was not solely imposed top down both because other people had knowledge that needed to be incorporated and because managers needed to get everyone focusing on the same goal). In changing the page on which they recorded these expenses, managers changed the way in which they wrote about them and achieved what they believed would be better control of these expenses. Record keeping in one way rather than another was a way of building managerial knowledge and control, not just recording what people already knew. As a matter-of-fact, an accountant at the meeting said that revising the way in which they budgeted might give them a "better identification" of what they were spending and where. The recording of expenditures didn't tell what managers already knew; it allowed them to know it in several different possible ways. Managers' decisions about how the budget should be written defined items differently; that is, it changed the nature of managerial knowledge.

The kind of knowledge any budget allowed was always partial and, to some degree, was even fictitious. This point was emphasized at a meeting to which I followed Doug immediately after the one I have just described. This second meeting was with his entire group

Strengths
Weaknesses
transformation

of team leaders and budgeting was also discussed there. When Doug described the budgeting changes that had been proposed at the previous meeting, his team leaders nodded approvingly. One of them said that the current system actually required "hiding" certain expenditures so that they could be spread out over several design groups in order to avoid an "unfair" assessment. For example, if an engineer ordered a test on a vehicle part and the vehicle's engine happened to break down during that test, the engineer's project could be charged for the engine repair, even though the engine had been used for ten years and after being repaired could be used for ten more. The engineering center avoided the unfairness of this assessment by spreading it out even when the budgeting system in current use did not allow for such distribution. In this case, the work of the engineering center slipped out from under the control of the budget, probably because engineers needed to believe in fairness if they were to be motivated to work toward the corporate goal. Carl G. Herndl (1996) has argued that employees sometimes resist dominant corporate discourse in order to uphold other values. For the engineers at Pacific Equipment, "fairness" would seem to be one of those values.

The engineers' resistance to the official budget requirements leads us to question the extent to which this text should be seen as coercive. It shows that subordinates have to cooperate in any construction of power. The fact that they could choose to subvert a written requirement demonstrates that when things run as planned, subordinates are contributing to that power construction rather than being overrun by their superiors. Thus their superiors have to persuade them to accept joint goals if symbolic power is to function.

At Pacific Equipment, the knowledge coded into the budget was less a reflection of some objective reality than an official version of knowledge that occupying a managerial position allowed corporate leaders to declare. Like other institutions, the function of Pacific Equipment is to enforce the definition of what counts as official knowledge, but presumably also like other institutions, it occasionally acknowledges that the official, unified version of reality can only approximate the more varied events that occur within its purview. Indeed, it is possible that such an acknowledgment is necessary if an organization is to succeed.

DOCUMENTATION

Along with budget, another important official genre in use at Pacific Equipment was documentation, by which management monitored subordinates' actions in distributing and amassing various forms of capital and by which the organization also continually created and maintained its own understanding of itself. The term *documentation* covers that genre of texts that makes official what has happened or, sometimes, what will happen (Winsor 1999). For instance, meeting minutes shape an organization's understanding of its day-to-day activity. Additionally minutes frequently name tasks that participants are publicly charged with performing and thus also shape an organization's understanding of its future. Note that minutes are not usually prepared by managers, but by people engaged in work at many levels. However, managers usually require documentation such as minutes and often enforce expectations for what will be in them even when other people do the actual writing (Winsor 1999).

Cheryl Geisler (2001) has commented on the extent to which the power of documentation rests in its ability to make actions public so that they can be monitored. In using documentation and requiring it from others, managers at Pacific Equipment are engaged in textually organizing work practices so that everyone works at establishing and maintaining the order that management has decided is most appropriate (Yates 1989).

Another way to put this is to say that managers use documentation to become what John Law calls "a centre of ordering." Law uses "ordering" rather than "order" to suggest the way in which any order is constantly under construction and maintenance. He argues that this construction and maintenance are controlled from a center point that renders people powerful. That is, power resides not in individuals, but in the positions they hold in a network of actors (cf. Foucault's 1980 understanding of power as relational). Law describes a center of ordering in terms of its ability to represent and thus establish knowledge about other actors in the network: "Roughly, then, a centre of ordering is (likely to be) a place which monitors a periphery, represents that periphery, and makes calculations about what to do next in part on the basis of those representations. . . ." (Law 1994, 104). These monitoring and planning actions are part of what generates the powerful position of managers. As I note in my

discussion of budgets, subordinates have to cooperate in any construction of order or power, and this cooperation can break down.
Managers use texts such as documentation to provide a constant
check on existing order and as an ongoing reminder to subordinates
of what is expected of them.

Documenting Engineers' Actions for Managers

Pacific Equipment managers monitored and planned the work of
others, using various texts as an important means to do so. Generally
speaking, engineers wrote reports and other texts only because managers required them to do so and they often complained about this.
(As chapter 3 will demonstrate, the engineers preferred to work with
more fragmentary texts showing data that facilitated joint interpretation.) While engineers and other subordinates such as lab technicians may have filled in the actual symbols in these texts, their form
was often determined by managers who institutionalized it either
through the creation of actual forms or through the expectations
they made clear to their subordinates. (Cf. Winsor 1999, 203–204.)
Thus even when engineers wrote about their own work, managers
often controlled the text and thus enforced a certain order.

For instance, at one meeting of Alan's group with their new
manager, the latter made several requests that the engineers adhere
more strictly to documentation forms that had been centrally established by corporate managers. Because the arrival of this new manager inevitably led to some changes, it provided a number of
opportunities where one could see management practices being
examined, and it was interesting to me that such changes were often
calls for closer adherence to documentation forms that management
had chosen. For instance, the manager announced that she wanted
the group to follow an official Product Development Process that
required filing written reports on all phases of the design process.
She said that when she had arrived at the engineering center, she
had asked for the "paperwork" on their current project "to bring
myself up to speed" and had been told that there were no such documents. The documentation was theoretically required corporate
practice; it required engineers to represent their work in a centrally
determined form, but the former manager of this group had evi-

dently not enforced the requirement and because engineers had
seen it as unessential to their own work, they had not been doing it.
That is, freed from the requirement to account for their work to
managers by means of an institutionalized genre, they had chosen to
spend their time in ways they found more interesting. The new
manager wanted this situation to change. She saw this documentation as a way to make the group's actions more accountable to the
overall corporate goals. That is, it was a text that fit the engineering field into the overall field of the for-profit corporation.

In another meeting, she called on her group to adopt a new
coding system for vehicle parts that would make it easier to compare
cost and performance of various components. As with the Product
Development Process documentation that was just mentioned,
engineers had not seen this coding as central to their own design
work and thus they had not used it. The manager needed to get
everyone focusing on the same corporate object, which she tried to
do by explaining to them why the changes were needed. In other
words, she tried to persuade them rather than rely on orders to shape *Balance*
their behavior. She needed their cooperation to maintain the hierarchy of corporate interests.

These examples of documentation all involve subordinates
representing their work to management in institutionalized forms.
That is, engineers might put the actual symbols on a page but there
is a sense in which these documents can be said to originate in management. Management's control of the text facilitated its control of
how people behaved, how they identified the goal of their actions,
and how they codified their own knowledge so that it could be used
by someone else for a purpose that differed from theirs. Control of
these texts was one way that management concerns were built into
engineers' actions and into the institution of the engineering
center. That is to say, it was a means to symbolic power, power that
is exercised by means of symbolic exchange rather than by any physical force (Bourdieu 1991).

Documenting Managers' Plans for Engineers

In addition to using documentation to monitor and organize the
work of subordinates, managers also organized work by disseminating

documentation such as organizational charts and mission statements to represent the order they wanted. This was a process that everyone seemed to understand and even welcome because it clarified their understanding of what was expected of them. They then could act in accord with the documentation's representation. For instance, I attended a meeting during which Alan's newly appointed manager met with her group of approximately twenty engineers for the first time. At the meeting, the new manager exhibited an organizational chart for the group. Alan later told me that he was glad to see this chart because he had previously not been sure whom he reported to at the interim level below the group's manager. The organizational chart was a representation that helped to shape how people behaved and to give control to managers in a manner that seemed more like guidance than force. That is, it was a form of symbolic power. It established the relative positions of people within the group. It also allowed the breakup of design over many people with central control (Bucciarelli 1988) allowing distributed cognition.

As Louis L. Bucciarelli emphasizes, such documents as organizational charts define reality as much as they describe it (1988, 92). Moreover, real work is usually less organized than the categories such a chart describes. For instance, at the meeting where this organizational chart was displayed, one attendee asked about which subgroup his work would fit into. Alan later told me that the questioner worked on parts that would fall into two different groups in the chart. Thus like the official budget I described that would theoretically charge test engine repair to a single design group, the tidiness of the chart is somewhat fictitious. Both budget and organizational chart are less descriptions of reality than ways of sorting reality so that it can be regulated. That is, they are documents that both flow from, and establish, positions of power.

During the meeting of Alan's group, the manager also displayed an overhead on which the group's goals were written. Pacific Equipment had recently built a number of factories outside of the United States and this engineering group was newly charged with supporting some of the factories. Thus the goals the manager exhibited differed from what the group was accustomed to seeing as its role. One engineer asked that the manager send a copy of the overhead to the whole group because it was the mission statement for which they would then be held accountable. The manager said that this statement was also what she had been given in her job description, simultaneously suggesting that when she showed the mission state-ment she was operating with the authority of the organization rather than with her personal authority (or, as Bourdieu 1991 might say, she had the institutionally granted right to speak) and that she, too, was accountable to the goals made official in a document.

Indeed, the people who functioned as "managers" within the engineering center all functioned as subordinates within the corpo-ration as a whole and had to document their own work for the scrutiny of others. So, for instance, Doug told me that the work of his Technical Services and Engineering Services group was evalu-ated by a series of "metrics," or measurable, recordable standards, such as the length of time between a design group's identification of a problem and his group's solution. Time to solution was measura-ble; it could be written down and moved around for scrutiny and assessment, or, as Latour says, it could be inscribed and function as an immutable mobile (1987, 227). Such immutable mobiles allow those in centers of power to scrutinize and organize the work of others and thus to fortify their powerful position. As Hutchins says about distributed cognition: "When the labor that is distributed is cognitive labor, the system involves the distribution of two kinds of cognitive labor: the cognition that is the task and the cognition that governs the coordination of the elements of the task" (1996, 176).

The "cognition that governs the coordination" of work is the knowledge claimed by managers, and documentation is one of the means by which managers lay claim to it. While documentation allowed managers to scrutinize the work of subordinates and to repre-sent the organization in desirable ways, it often allowed managers to create the impression that the required form or the mission statement had somehow originated outside of their control, creating the sense that management itself always existed elsewhere. Alan's new man-ager, for instance, called on her subordinates to use documentation that she herself had not designed. The constant removal of manage-ment also occasionally allowed the managers within the engineering center to claim solidarity with their own subordinates as Alan's new manager did when she said that the mission statement had been given to her. Thus it facilitated the concealment of symbolic power that Bourdieu (1991) says is necessary for its effective use. If I had been able to observe at Pacific Equipment's headquarters, I might have seen a point where responsibility for power was accepted, but equally, power might always conceal itself, and the presumption that it exists openly somewhere might be a delusion. People at headquarters might attribute their actions to pressures from the stockholders or the government or market forces. At any rate, the operation of power was often concealed at the engineering center. Insitutionalized genres such as budgets and documentation seem to exist in an autonomous form with no one claiming authorship.

Agency

(NOT) DOCUMENTING MANAGERS' ACTIONS

In addition to requiring that their staff submit textual representations of their work for managers to use, managers also tried to create textual representations of management work for others to use. They were concerned about the staff's attitudes and perceptions and worked to affect them because they wanted and, indeed, needed, employees to all work toward more or less the same goal. Thus, managers tried to control both the texts that represented the work of their subordinates and the texts that represented their own work.

As this discussion suggests, managers at Pacific Equipment's engineering center do not rely solely on issuing orders to achieve organizational goals. Their reluctance to rely on orders suggests their tacit recognition that power is relational. It's granted. Bourdieu argues that symbolic power can exist only because the person submitting to it trusts and gives credit to the authority (1991, 192). Because power is relational, managers had to work to maintain the cooperation of the engineers in sustaining their own positions. One engineer, Alan, told me that managers who worked closely with engineers particularly needed to explain decisions. He said that engineers were more likely to accept unexplained decisions from higher levels of managers because engineers believe management's assertion that a certain amount of confidential planning is necessary (another reason for displacing management decisions to someone somewhere else because such decisions are less likely to be questioned). But the closer managers were to engineers, the more they had to concern themselves with engineers' reactions to decisions and plans. That is, they needed engineers' cooperation in order to maintain their own positions. Managers at the engineering center seemed well aware of this need. However, while they knew that they sometimes needed to justify their actions, managers also demonstrated recognition that documentation allowed for scrutiny of actions, and they did not always believe that such scrutiny of their own actions by subordinates was desirable. Thus they occasionally avoided documentation altogether.

The highest-level manager I observed was Ken. There's a certain irony in the fact that he was also the person I observed who was most openly concerned with strategizing about how to get various groups to accept management's plans and decisions. One of the most interesting meetings I observed involved upper-manager Ken, a midlevel manager who worked for him, and another one of my participants Brad, who worked for the midlevel manager. Brad was officially a team leader for advanced technology development but he currently had no group. He had been with Pacific Equipment for only six months. Ken had evidently given Brad six months in which to establish himself at the engineering center and to examine the possibilities for use of advanced sensor technology in off-highway equipment.

The purpose of the meeting was to decide how to structure Brad's work so that his coworkers would accept the new technology he was developing. At the meeting, Ken said that he believed that separate advanced technology groups were "dangerous" because they were isolated from the rest of the organization. He said that no one would listen to such a group because it was not "part of the pack." Thus he did not think Brad should be given a separate advanced technology group to supervise. On the other hand, he also did not believe that Brad should simply be incorporated into an existing group because such groups tended to focus on shorter-term needs and did not look beyond the next one to two years. Eventually, the three men agreed that Brad would work on a project that would fulfill a product design group's need for advanced technology. Ken told Brad to choose a project with the highest probability of success. The object would be for Brad to show what advanced technology could do for the product and thus provide what Ken called "a practical example of how we bring some value to the party."

This discussion offers interesting suggestions about how knowledge and power are created and accepted in a technical organization. Information does not just "flow" from the research to

justification

the product groups. There is resistance along the way because of possible upset to the organizational structures that have already been established. Old-timers can be threatened by newcomers; using new technology, new knowledge, can mean that someone's organizational role is rendered obsolete. That is, we again see an example of knowledge structures and social structures that overlap and influence one another. Note that even this highly placed manager was not able to mandate a change to these structures. Ken did not seem to feel that he could impose knowledge about sensors through his managerial authority. He believed that knowledge had to be insinuated into existing groups and practices. A person bringing the information also had to ingratiate herself so that people would seek her out and listen to her.[2] As Barry Smart says about Michel Foucault's idea of power, ". . . power is not conceived as a property or possession of a dominant class, state, or sovereign but as a strategy" (1985, 77). Ken's notion of how to exercise his position of power was highly strategic. Power was a condition he had to work to generate rather than a possession upon which he could simply draw.

In contrast to the practices I have described, Ken's strategizing was also a rhetorical practice that was largely untextualized, at least in a traditional sense. Rather, it functioned through the distribution of people and tasks. It was less written on screen and paper than on the shape of the engineering center. I suspect that this difference originates in the capacity of writing to make actions and ideas visible so that they can be scrutinized. Managers thus find writing particularly useful in regulating the actions of subordinates, or, perhaps, in representing their own actions and goals so that subordinates can be enlisted in approving them. As I pointed out, Bourdieu argues that symbolic power, what he defines as "the exercise of power through symbolic exchange" (Thompson 1991, 23), can exist only because those who submit to it trust and give credit to the authority (Bourdieu 1991, 192). However, Bourdieu also argues that symbolic power is most effective when it is least visible. As Bourdieu says, "symbolic power is that invisible power which can be exercised only with the complicity of those who do not want to know that they are subject to it or even that they themselves exercise it" (164). Ken often wanted his persuasive management of work in the engineering center to remain invisible and unexamined by subordinates, evidently because he, like Bourdieu, believed that such unscrutinized actions were sometimes more effective. Thus he may have been less likely to write about those actions. Bernadette Longo (2000) has argued that technical writing is one means by which authorities exercise what Foucault (1979) calls a "normalizing gaze." Such a regulatory scrutiny of engineers by managers is, for instance, the function of the documentation I have just described. In hierarchical organizations, however, there is usually no textual mechanism that allows this gaze to be reversed; engineers do not officially scrutinize the actions of managers, although, as I will show in chapter 5, they do scrutinize those of laboratory technicians and, as I suggested, the actions of the engineering center managers are scrutinized by managers at corporate headquarters. The regulatory gaze is always directed downward, and thus documentation is typically required of subordinates by their superiors.

REPRESENTING AND REGULATING ACTIONS

Based on this discussion, we might suspect that in organizations like the engineering center, texts make actions visible and thus available for scrutiny and regulation. Managers are able to organize activity partly through institutionalizing genres such as documentation or budget categories that name and organize subordinates' actions. The creation and organization of these texts is thus simultaneously a creation and organization of managerial knowledge. In turn, this managerial knowledge organizes and reinforces positions of power within the corporate field that envelopes and makes use of the engineering field within it. In this system of power and knowledge, we see an exemplification of the sense of power that Smart attributes to Foucault:

> . . . an increasing inter-relationship between the exercise of power and the formation of knowledge which followed from the disciplinary transformation of institutions into apparatuses within which methods for the formation and accumulation of knowledge began to be employed as instruments of domination and increases in power began to produce additions to knowledge. (1985, 91)

To represent something is to gain knowledge of it, which in turn means gaining a degree of power. Creating representations is a means to knowledge both of objects, such as off-highway equipment, and of people's actions, such as the way in which they spend their time. Indeed, creating representations is one way to define what counts as knowledge and thus is itself an exercise of power. In turn, knowing objects and actions generates the means to control them. Thus one technique that managers can use to generate positions of power for themselves is to control both what representations are created within an organization and the shape those representations take. Some actions, such as Ken's strategic moves, can remain unrepresented and thus remain closed to undesired scrutiny. Other actions, such as the way in which test costs are distributed, can be represented only in official ways that ignore deviations but that support managers' right to name activity as they wish. Still other actions, such as engineering activity, can be represented in forms such as budgets that are useful to managers but foreign to engineers (so that, in vignette 2, a team leader was unaware of how his favorite project was budgeted). Through all these means, managers structure both the knowledge and distribution of various capitals within the Pacific Equipment engineering center.

At Pacific Equipment, then, management was not a brute assertion of power. Rather, managers worked at least partly through controlling representations: they created representations of their own decisions and plans that were addressed to workers, and they required workers to create regulated representations of their activities that managers could use in decision making. Regulating through institutionalized representations meant that they were able to embed management concerns into the culture and practices of the organization itself so that most members engaged in carrying them out even without direct supervision from a manager. Managers' ability to control these representations both grew from, and led to, their ability to convince employees and their ability to organize them. In other words, effective use of textual representations allowed managers to exercise symbolic power within the engineering center. Such power was not permanent but rather was always under construction with institutionalized genres as one means leading toward this goal. That is, power was jointly generated by the more and less

weekly updates

powerful as they coordinated their actions around official genres that represented what happened or should happen in the engineering center.

What this chapter shows is that for Pacific Equipment managers, budgets, schedules, and documentation were important genres that allowed them to channel engineering activity in ways that would lead to corporate profit. This does not mean, however, that they were the only important genres in the corporation. It is possible to argue that they weren't even the most important, because while they did work that was useful to the for-profit corporation, other work, requiring other genres, was done in other parts of the engineering center. Chapter 3 will show that for engineers, the most important capital is the cultural capital represented by interpreting technical data. This work was essential for the corporation's success and it operated through different genres, the most important one being the data report or instrument trace generated through laboratory work. In vignette 3, we will see a group of engineers working together to try to generate the cultural capital of engineering knowledge from the raw material of data.

Vignette 3

A Meeting with Engineers: John

As I reach the engineering center a few minutes before 9:00 A.M., I realize that I have forgotten my briefcase. Fortunately, when John comes to fetch me from the security desk and I tell him about my lapse, he takes me immediately to a supply room and gives me a legal pad and pen. We go to John's cubicle until it is time to go to this morning's meeting. John has previously told me that the subject of the meeting will be changes that need to be made to a vehicle's transmission in order to improve its performance. He says that he and other engineers have ideas for how to achieve their goal but all of the ideas would require money and design time and test time and thus it is not easy to get management approval.

In John's cubicle, he generates several overheads for the meeting. When he is finished, we go to one of the meeting rooms that are lined up along the center of the floor where meetings can take place with some privacy. We are the first ones there, and John discovers that the room has no overhead projector in it. He goes off in search of one. While he is gone, Sam, a design engineer, comes to the door and hesitates when he sees me, evidently assuming he is in the wrong place. I introduce myself and he sits down, gets out a notebook, and starts to write. Sam has been newly assigned to do rear axle design on the vehicle. If the people at the meeting decide that the rear axle has to change (which is possible), he will have to figure out how to accomplish the desired change. John comes back with the overhead.

Derek, John's team leader, arrives. John begins to introduce me, but Derek says that he saw me at a meeting of managers when I was shadowing Ken, the upper-level manager. Like Sam, he immediately opens a notebook. Two people are still missing, but John starts the meeting a few minutes after 9:30. He begins by showing an overhead designed with five bullets that represent the objectives he has set for the meeting. The first item on his list is a testing plan. Derek says that they now have a test cell and technician assigned to them, but they need to create a plan for what to test. On an overhead appears some data from previous testing. He says that they know that a change in axle affects performance in some way but they are not certain exactly how. They don't know "enough."

Next, John shows another graph on which he has extrapolated from data on the power of current vehicles to what will happen if Pacific Equipment makes changes that are being contemplated. He says that he doesn't know if the extrapolations are fair. Derek says that the graph doesn't agree with what they saw in a different graph. He pulls another graph from the pile of paper that John has brought to the meeting. They all look at it as he talks. John suggests that the data might vary because they were taken with or without pumps attached, for instance, and Derek agrees. John says he had trouble interpreting the data because they came on a spreadsheet "with more data than you could ever manage." He offers to "play" with the data some more.

They continue to talk about whether the graph "makes sense." "We ought to get more data," John says. In other words, they need a good test plan. John says it's hard to implement plans when two people are still missing, although, he jokes, they could make assignments anyway. The three men laugh. John offers to meet with Matt, who is one of the missing meeting attendees and is in charge of testing drivetrain parts, to develop a test plan. He admits that it may be difficult to get everyone together at the same time.

Derek says that they need to establish goals, starting with the customers' needs but they don't know exactly what those are.[1] He says they do know that customers prefer a rival company's vehicle in the area of performance that they are concerned about. John says that this rival company did a good job of managing its vehicle's oil flow. "When we took it apart," he says admiringly, "there was very little oil floating around." At this point, Duncan arrives. He is in charge of front axle design. He apologizes for being late but says he was dealing with the "crisis du jour." A few minutes later, Matt arrives.

They continue to talk about goals for the new vehicle. John says that they can set their goals for this new vehicle in a number of ways: pull them out of the air, base them on testing, or base them on customer needs. He says that the last would be ideal but they don't know what those needs are. Duncan describes the goals that his group has established for the front axle.

Derek raises the issue of cooling. John says he talked to the engineers working on a different vehicle about their experience with overheating. Matt says that the engineers in charge of cooling will claim that cooling will not be a problem, but Derek is suspicious. Matt says that in his opinion, they need a radical redesign of the drivetrain in this vehicle, not incremental improvements. Duncan adds that they have already pushed incremental changes as far as they can without adding cost. Matt argues that it is better to get rid of the problem right from the start of the design because that will meet the customers' needs better. Functioning as a manager, Derek says that would be ideal but there are schedule and cost problems. He says they'll need more data before they decide.

John says that they need to itemize where their power loss is coming from. Derek proposes that John should set some goals for the various component areas and then meet with people from those areas. He says that everybody may hate the goals John has created, but it they all hate them equally, then John has probably done something right. John agrees to do this.

John asks how the cost for this project will be allocated. Derek says that the project will be funded by a separate account. He says that Ken has agreed not to set a cost target until they have more data. John mildly asserts that they are going to need money to accomplish anything. Derek says that once they have some numbers together, they can go to management.

The meeting ends around eleven. As we are leaving the meeting room, Matt asks me how John managed to get someone to take notes for him at his meeting. John laughs and introduces me. As John walks me out of the building, he tells me that he wanted goals to be set at this meeting and that did not happen. He strategizes a little about how to accomplish his goal at the next meeting.

CHAPTER 3

Negotiating Knowledge Across, Down, and Up the Hierarchy

In God we trust; all others bring data.
—Sign posted on the cubicle of a
Pacific Equipment engineer

Wherever computations are distributed across social organization,
computational dependencies are also social dependencies.
—Edwin Hutchins, *Cognition in the Wild*

As chapter 2 demonstrated, Pacific Equipment managers balance budgets of different kinds of capital and are able to include engineers' technical knowledge in this budgeting activity because the cultural capital represented by that knowledge is translated into other forms with which managers can work. The cubicle sign I just quoted can be seen as a tacit recognition of an engineering equivalence between data about the designed object, the basis of engineering knowledge, and the "cash" that is valued in both the for-profit corporation and the more common version of the saying: "In God We Trust; All Others Bring Cash." In this chapter, I move from the field of the for-profit corporation to that of engineering, where the cultural capital of technical knowledge is most highly valued. This chapter will examine how engineers communicate up, down, and across the organizational hierarchy in order to persuade others to accept their notions of what counts as accurate knowledge, and thus to gain some measure of control over knowledge production at the Pacific Equipment engineering facility. It appears that generating knowledge entails generating power and that rhetorical action is a means to both. As Michel Foucault says, "Far from preventing knowledge, power produces it" (1980, 59). Power over something or someone allows us to scrutinize it more easily and thus to build knowledge of it. Being in a position of power also allows us to claim authority for our understanding of what we see.

In Pierre Bourdieu's terms, a field's members always struggle to control the capital that matters to them. Indeed, according to his editor, John B. Thompson, Bourdieu defines a field as a "structured space of positions in which the positions and their interrelations are determined by the distribution of different kinds of resources or 'capital'" (Thompson 1991, 14). However, while Bourdieu is interested in how capital is distributed, a process that he sees as inevitably competitive, he does not talk much about how cultural capital is generated, an action that is usually more cooperative, at least in engineering. I am sure that Pacific Equipment engineers experienced some sense of competition over the control of knowledge. I did see discussions in which people at least began from points of disagreement and engineers spoke up for their own ideas and opposed those of others. Indeed, vignette 3 might serve as an example of such disagreement. However, in the discussions I saw, competition tended to be downplayed as they worked toward a common goal: the creation of a quality object. Such an object is the professional and usually personal goal of engineers, and, from a management perspective, the company needs this in order to generate other capital. Both engineers and managers will evaluate the engineers' work by the working of the product (although, of course, what counts as a "working" product is also a matter for negotiation). Engineers might have disagreed about the best way to create a quality object but, because they shared this goal, they were usually able to negotiate. Thus, among the engineers at Pacific Equipment, any struggle for control of cultural capital was on a small scale. As I will show, its daily and hourly enactment was accomplished in acts involving rhetoric, or the persuasive and creative use of language. As I will also show, the generation of the cultural capital of knowledge also seemed to be at least partly achieved by rhetorical means, including texts generated by test instruments and by oral negotiation around those instrument traces.

The work of the engineers largely consisted of planning for how the lab technicians would create these texts (an action the engineers did not see as persuasive) and then analyzing and interpreting the texts they had amassed. The work of creating the texts was thus distributed between engineers and technicians, while the work of interpretation was usually distributed among several engineers. Through jointly endorsed interpretations, these texts came to represent "facts," and although my observations showed that technical facts were not self-evident and agreement about their nature was not automatic, any persuasion operating here tended to be transparent to those involved.

In contrast, managerial knowledge and consequent decision making seemed to engineers to be more open to persuasion. That is, when "facts" were transferred across fields, their relevance and stability became less self-evident. Managers needed to be persuaded that engineers knew what they were talking about and should be allowed to continue in the direction they wanted or should be given the resources they had requested. Craig J. Hansen describes employees using texts for similar purposes in an organization he observed:

> In the specific setting of my study, writing served as a means to claim and maintain authority. Authority, in this context, does not necessarily mean the right to direct the work of others, but rather, to self-direct, to minimize unwelcome managerial interference, to maintain a position in the formal and informal hierarchy. (1995, 105)

Thus when addressed to managers, engineers' texts often became a tool to try to appropriate some decision-making power although the need for the text to be conventionally persuasive also recognized that the engineer occupied a less powerful position than the manager. In this chapter, I discuss how the engineers used their professional (and I would say rhetorical) skill in generating textualized "facts" about objects in order to negotiate knowledge within the field of engineering and how they used somewhat different rhetorical means to try to preserve their control of engineering knowledge in their communication with managers.

The participants I will concentrate on in this chapter are Alan, Dan, Greg, and Paul. Alan is an engineer who designs new vehicles. I observed him in the summer of 2000, when I was focusing primarily on the interaction between engineers and managers. I observed

the other three men in the summer of 1996, when I was most interested in whether engineers saw their writing as persuasive. However,
all four sets of observations yielded material that is relevant to discussions of power, knowledge, and text. Dan is a mechanical engineer who primarily conducts testing of vehicle drivetrains. Greg is a
materials engineer who serves as a consultant to Pacific Equipment
engineering groups who want to use any sort of "compliant" material (i.e., material such as rubber that can be deformed without
losing its properties) in the part of the vehicle for which they are
responsible. Finally, Paul is an electrical engineer who functions as
a team leader for several engineers working on electronic controllers
used in Pacific Equipment's vehicles. His role as team leader meant
that he spent less time than the other men in generating data and
more time keeping track of the conclusions that his group had
reached about the data they had generated. They spent their time
turning objects into texts; he spent his time keeping texts ordered
and accountable to management.

In analyzing my observations, I draw heavily on Bruno Latour's
work on inscription, or what Latour calls "immutable mobiles"
(1990, 26). Latour talks about the process by which material reality
is converted into written traces by means of instruments such as
mass spectrometers that produce paper objects on which reality has
been written down or "inscribed" (22). Latour says that these paper
objects are more useful for analysis than the material objects they
are based on because, once created, inscriptions remain fixed,
whereas material objects can vary alarmingly depending on such
variables as ambient temperature or the way in which the sample for
testing has been prepared. Additionally, these paper objects are
mobile; that is, they can be moved about from office to office and
even from continent to continent and thus can be massed together
to create powerful bases for knowledge. In other words, through
inscriptions, power and knowledge are coproduced. The engineers I
observed produced writing that could in part be described as inscriptions. These inscriptions allowed them to build joint knowledge of
the machines they studied and thus to control them. The engineers
also tried to use texts built on these inscriptions to maintain control
over their own work when managers seemed to be threatening to
interfere in what engineers thought of as their quality-based portion
of the engineering center work.

KNOWING AND CONTROLLING OBJECTS

Engineers typically work with material objects, but much of this work is actually accomplished in a manner that is mediated by texts. For instance, the design or development engineer first works with an object that does not yet exist except on paper. (See Bucciarelli 1994 for a sustained example of this work. Also cf. Medway 1996 on architects.) Paper forms of the object are suggested, varied, and negotiated in the corporate settings in which engineers work. Engineers' knowledge first flows from these paper forms to the object but then must come from it, as they seek to understand and modify the actual workings of what they have created. (See Latour 1996, for an example of this back-and-forth movement.) Even after a technical object exists, documentation about it can serve as a more reliable form of the object than the object itself. Thus, for example, Paul said,

> if there is a question about the way the software works, then we pull out the specification and we look at it and we say, "Well, OK this is what it says. This is what the software does. What's right?" And if they're different, we have to decide if the spec is right or if the software is right, and then we change the one that's not right. So the importance of having the documentation right is that it's the only thing that tells us in English what this thing is supposed to be doing.

As Paul's comment shows, it was not always apparent whether the paper (the "spec") or the artifactual version of the object was the "real" one. Such texts were the means by which engineers created, scrutinized, and gained control over the objects upon which they worked. These texts included engineering drawings and specifications that describe the object, work orders that engineers use to instruct technicians to build or test the object, and test reports that technicians return to engineers for analysis.

Thus, although we normally think of engineers as working with objects, my observations suggested that any knowledge they gained from objects had to be supplemented by knowledge gained from texts. This was true despite the fact that engineers pride themselves

Fig. 3.1. *Dan's Office*

on their direct contact with objects and on their apparently unmediated understanding of them. The engineers I observed were no exception. For instance, Dan told me that he spent about half of his time in the lab and test area that constituted about two thirds of the Pacific Equipment engineering facility. One of the tasks Dan regularly performed in the lab was to check on long-running tests. He said that such actual observation gave him a "feel" for what was happening: "I've listened to the test stand long enough and know enough about it that when it does something that isn't right, I can spot it." Here, Dan described using tacit, sensory knowledge about the devices he was responsible for. As I will show, such unmediated knowledge about a machine is seldom possible, but engineers value it highly and being able to claim access to it is one of the ways in which they place themselves in powerful argumentative positions.

The engineers' offices also illustrate the extent to which they try to draw knowledge directly from objects. For instance, Dan's office (shown in figure 3.1) contains files and notebooks filled with paperwork, posted cartoons, personal memorabilia, and a computer that he uses primarily for word processing and E-mail. To some

power
kinetic
tactile

Fig. 3.2. Dan's Desk Exhibits a Mix of Gears and Paperwork

degree, it is indistinguishable from the office of any working professional person. However, on his desk (barely visible in the far right-hand corner of figure 3.1) is a clue to the fact that an engineer works here. (That part of Dan's desk is shown in more detail in figure 3.2.) Among the paperwork there, we see a collection of gears that have exhibited problems Dan is working to solve. He uses these gears partly as visual evidence. He told me that "if someone says the tooth can't look that bad, I can show them the part and say 'here it is,'" a clear example of an object being used to generate knowledge, although I would argue that this knowledge comes less from an unmediated contact with the gear than from Dan's rhetorical framing of it as evidence. We also see material objects in Greg's office (shown in figures 3.3 and 3.4). As figure 3.4 shows, on the floor behind Greg's desk, seals and other parts that he worked on lay adjacent to the documentation about them.

However, although these engineers' offices do indicate their occupants' direct contact with material objects, they also illustrate another point: The objects are present in paper as well as (and sometimes instead of) artifactual form. The vehicle will not fit in

*visual
fountain*

Fig. 3.3. *Greg's Office*

Dan's office in its material embodiment; however, it is present there in the documentation. Indeed, the ability to create and manipulate paper forms of the object is what distinguished the work of the engineer from the work of the technician at Pacific Equipment. And while there were gears on the floor behind Greg's desk, figure 3.3 shows that that desk was buried in paperwork of almost academic proportions. Additionally, behind his desk were rows of reference sources that he used when assisting various Pacific Equipment groups to select appropriate material for seals and gaskets. This mixing of paperwork and artifact is typical of engineers' offices, showing the extent to which texts are needed to supplement any direct knowledge of an object.

Paper forms of objects are highly valued because they possess a flexibility that the objects themselves do not have. By means of writing, objects can expand and contract in space. They can travel over the world in documents and drawings. During the time I was an observer at Pacific Equipment, for instance, Dan was sending a vehicle's drivetrain to Germany, a process that took a great deal of time and trouble. However, he also talked about communicating with German engineers via E-mailed reports to which he attached

supplement

Fig. 3.4. *The Floor of Greg's Office*

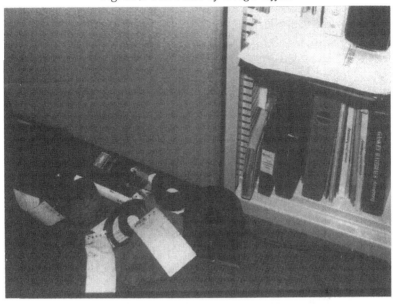

graphs showing a coefficient of friction curve. In the substitute form of those reports and especially in those graphs, the drivetrain was present and traveled much more efficiently than it did in its artifactual form.

Documents also serve as memory extenders. Such memory extension was the primary function of the notebooks in all four offices. "When you have a technical specification," said Paul, "you don't remember for each thing what all the properties of that thing are." Written documentation helped to solve that problem. Thus, I observed Dan leafing through files of papers as he tried to answer a question from someone on the phone. Documentation, Dan said, was "a way of retrieving things when later on you try to look up to see why that change was made in the part or when it was changed." Greg spoke similarly about engineering drawings. "Without the engineering drawing," he said, "your eyes and your hands are not accurate enough to see subtle changes. And also your memory is not good enough to carry forth what you saw yesterday to two weeks or three years later." We tend to think of an object like a vehicle as stable and unchanging, but to the engineer who works on it, a vehicle is always changing rather than remaining fixed. Indeed, the development engineer's job is to ensure that the vehicle changes in ways that will give it a continuing advantage in the for-profit marketplace in which the corporation operates. Thus, these engineers study a moving target. Writing, however, functions as an "immutable mobile" (Latour 1990, 26), mobile because it can be moved about in space and time but immutable because even when it is moved about, it retains the same characteristics and exhibits the same data. Thus, engineers accomplish their work partly by freezing the moving object into a written form they can study. Test reports and technical documentation are thus tremendously powerful texts for engineers trying to gain knowledge and consequentially control of an object.

USING ENGINEERING TEXTS TO
UNDERSTAND THE OBJECT

These engineers, then, used texts to increase their understanding of the objects they worked with. Although some understanding can come, as Dan says, from such activities as listening to an engine in the lab, such means of understanding are never seen as sufficient (probably because they can be neither shared nor amassed) and are always supplemented by examinations of textual representations of the object. In engineering, the transition from object to text is usually accomplished in several steps, the first of which is the use of test instrumentation. Instruments produce textual representations of the object that are fragmentary rather than sustained prose, but because these representations are seen as closest to the object, they are very highly valued and give an engineer powerful evidentiary material. For example, one instrument that Dan commonly worked with was the strip-chart recorder (shown in figure 3.5). To use a strip-chart recorder, a technician hooks probes to the engine or to other mechanisms being tested (the transmission hooked to probes for this recorder is shown in figure 3.6), and the recorder traces the outcome directly on paper. The recorder mediates between the transmission in the lab (shown in figure 3.6) and the paperwork in Dan's office (shown in figure 3.1), turning one into the other. It is thus an instance of what Latour says is instrumentation designed to create the illusion of nature "writing itself down" (1987, 70) and is valued

Fig. 3.5. Test Stand Control Room

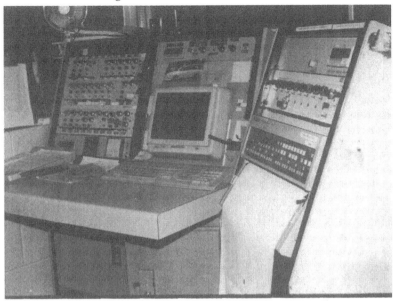

Fig. 3.6. Test Stand Holding Pacific Equipment Transmission

partly because of the implications of that illusion. Specifically, although human beings have designed the instrumentation and test procedures, the test results are recorded without human interference; the data thus appears to have come into existence untouched by human hands. The possibility of such "pure" data in turn suggests that one can know the machine directly, without an intervening screen of human perceptions and prejudices. Within engineering, such data are very powerful.

At an extreme, it is even possible to treat these instrument traces as writing done by the machine to the engineer. Of the four engineers I mention in this chapter, Paul, who worked with computers, was particularly likely to talk in these terms. Thus, as he programmed an electronic device called a "controller," he spoke about "communicating" with it: "Sometimes I'm talking to it and sometimes I'm listening to it. This [command] tells it to be quiet. Something is hung up because it doesn't like to be quiet because nothing is happening. Here it said 'OK, I'll be quiet.' . . . It's trying to talk to me and it's not doing it very well." In our interview, I asked Paul if this anthropomorphic talk was for my benefit or if it represented the way he really thought about the controller. He said that he thought of himself as communicating with the controller because the controller was responsive. In the coded messages it sent to Paul's terminal, it wrote to him.

> Paul: It's not just sitting there doing one thing. It is actu-
>   ally responding to what I send it.
> Dorothy: And because it's responsive like that you think
>   of it as—
> Paul: A responsive, decision-making machine.
> Dorothy: Interacting with you.
> Paul: Yes, it is. And we put in the logic so it can do that.

Of course, Paul is quite clear that the computer's capacity to react comes from the logic with which humans supply it. Even given this caveat, however, he saw the machine speaking for itself through code, as though the machine could represent itself directly.

Such a treatment of instrument traces ignores the degree to which they are humanly shaped and, even more, the degree to which they must be humanly interpreted for knowledge to be created. The problematic nature of uninterpreted data is evident in vignette 3, where John says that he currently has "more data than you could ever manage," but the disorderly nature of this data leaves him and his colleagues unable to understand the phenomenon at hand. The paper results of testing do not constitute knowledge in and of themselves. Rather, these documents must now be interpreted, and, in order to count as knowledge, the interpretation must often be jointly arrived at by more than one person, as meeting attendees try to jointly do in vignette 3. Indeed, my observations suggest that the joint interpretation of instrument traces through oral discussion is one of the chief ways in which these engineers built consensus among themselves and thus created what was accepted as engineering knowledge (cf. Mangrum, Fairley, and Wieder 2001).

For instance, I observed Dan and his coworker Tom examining a strip-chart together. Their following conversation shows that they are engaging in what K. Amann and Karin Knorr-Cetina (1990) call "the fixation of visual evidence," that is, the communal discussion of visual evidence to come to a common interpretation of its meaning:

> Tom: Which one actually goes to the pilot valve? [Dan points out the trace Tom wants.]
> Tom: Here. You see right here. Forward starts out slow here. And here is the neutral.
> Dan: It's slow, too.
> Tom: But look how much faster it jumps up. [He theorizes about why the trace looks like it does. He asks Dan if there is a trace taken at a higher temperature, and Dan points out one that would have been taken after the oil warmed up.] See, and this one never dropped. That's probably why it's working as well as it is.

Note that Dan and Tom use the instrument trace to jointly construct an understanding of the vehicle that they could not gain directly from the vehicle itself or even from the strip chart without one another's input.

I observed a similar occurrence between Greg and four coworkers at a meeting. The five engineers spent most of the meeting focused on a set of instrument traces called "real-time pressure" distributions and on an engineering drawing. The real-time pressure distributions were pictures in which different colored areas represented different pressures exerted by a roller on a pressure mat. The meeting began with Mary, the engineer who had created the distributions, laying out two sets of these pictures on the table around which they all sat. One set of pictures showed the results from a product being developed by Pacific Equipment, and the other set showed the results from a product made by a rival company. The engineers first discussed the pictures in order to come to agreement about what they meant:

> Rob: [pointing to one picture] So this for sure *tells us* . . . Somehow we have to make our prints look like this [pointing to the rival company's pictures].
> Peter: [pointing to three pictures in succession] And this and this and this *shows* the transition.
> Greg: It looks like this side is taking all the load. . . .
> Mary: What would be useful is figuring out. . . .
> Rob: What I was after, *the pictures will show.* (my emphasis)

Shortly after this, Rob got out a set of engineering drawings, and the five engineers began looking at them.

> Rob: I think this *tells* a lot about how this . . . is connected. . . . It sure says a lot to me.
> Mary: That's pretty *telling.*
> Jim: What's the explanation for? . . .
> Rob: [pointing to the drawing] It's this. (my emphasis)

Note that, like Dan and Tom, Greg and his coworkers treat the instrument traces and engineering drawing as texts to be interpreted. As Mary says, they are something that can be used to "figure out" something "useful." Note, too, that the texts are convenient substitutes for the vehicle. For instance, at the beginning of this excerpt, when Rob wants to talk about changing the vehicle, he talks about changing the pictures as if it were the same thing. When he wants to explain the vehicle's behavior, he points to the picture. The traces and drawings represent the current state of the vehicle and thus can be substituted for the actual vehicle in discussion. This identification of the text with the object is not surprising because the pressure distribution pictures and the engineering drawing bring the vehicle into the meeting room, where its size prevents it from being present in its artifactual form. But the traces also go beyond the vehicle; they "tell" the engineers something that the vehicle itself does not. They provide information about the vehicle that cannot be obtained by looking directly at it. However, they provide this information only when engineers interpret them, a process that is most frequently communal. Thus, even an engineering drawing is not simply a duplication of the object. As a text, it is open to interpretation in the way the object itself is not because it is more easily available and meaningful in the sense of "full of meaning" that can be publicly interpreted. From the perspective of rhetoricians, engineers' interpretation of instrument traces is, like all interpretation, uncertain and thus is always rhetorically achieved.

In our interview, Greg spoke about the way in which the pressure distribution pictures served as an occasion for him and for his coworkers to negotiate a common understanding of the object they were studying. First, he said, he and his coworkers had "to explain what we were getting. . . . Explain what is physically happening in Newtonian physics, if you will, as well as explain to each other, to understand, to establish a base of communication. . . . We have to establish a common ground. . . . We explained what we thought the data meant." In addition to telling the engineers something that the vehicle itself did not, the pictures provided a way for the engineers to come to a rhetorically achieved common agreement, just as Dan and Tom came to a common agreement about what they saw in the strip chart.

I also saw engineers use the creation of an engineering drawing as an occasion to negotiate over the design of a future product. When Alan asked one of his coworkers about a throttle mechanism the coworker was designing, the two of them sat together to look at design sketches the coworker had made. Looking at the sketch created an occasion for collaboration. They jointly modified the drawing as they talked through various issues, both of them drawing on the same sketch, an event I mentioned in vignette 1. The engineers' talk both grew from, and led back to, the drawing. Alan used the sketch as evidence that in the current design a worker would have a hard time getting at a nut with a wrench, and then the coworker showed him the drawing to explain why one solution that Alan proposed would not work. Eventually, Alan proposed a different solution and they modified the drawing by ripping a second sheet of paper into a curved shape and holding this shape up to the drawing so that they could visualize better what was happening in three dimensions. This modification seemed to satisfy them both. Like instrument traces, drawings are texts that are particularly important in engineering and that engineers seem to use both as occasion for negotiation and as evidence in the negotiation process. (See Henderson 1999 for a discussion of the importance of drawing in engineering.) These texts are part of the way distributed cognition is effected in engineering in order to create knowledge. Engineers' negotiation around them provides us with one answer to what David Middleton calls a "critical issue": "how to conceptualize team expertise and practice as accomplished in the social actions of team members in dialogue with each other rather than as some bureaucratic summation of individual expertise and how to study teamwork as the accomplishment of 'dialogical' rather than 'monological' expertise" (1996, 234).

Drawings and instrument traces are, then, writing that is very closely tied to objects and thus are powerful texts in engineering. However, these texts' meaning is not determined by those objects. Engineers build consensus within their own work group through orally negotiating the interpretations of texts that seem to come almost directly from the machine itself. In so doing, they gain control over the machine and also shape one another's actions. In vignette 3, when John and his colleagues eventually agree upon what their data mean, such agreement will shape how they decide to modify the vehicle. Shaping actions may not be precisely synonymous with "controlling" those actions but such negotiation around texts is certainly one of the means by which power is appropriated and distributed among engineers in the Pacific Equipment engineering center.

USING TEXTS TO NEGOTIATE POWER WITHIN THE HIERARCHY

While instrument traces helped engineers to negotiate knowledge with their colleagues, other texts became necessary if they were to draw on the knowledge of the laboratory technicians and then control interpretation of their work as it passed out of engineering and into the hands of managers. The aeronautical engineer Walter Vincenti says that because most engineers work in corporations, engineering is normally done in overlapping intellectual and organizational hierarchies. That is, engineering problems are normally ill-formed problems at the management level. Managers define these problems to some degree and parcel them out as assignments to various engineering departments. Within those departments, supervising engineers define the problems still more concretely and assign them to individual engineers or engineering groups. This structure affects both the kind of thinking and the kind of writing engineers need (and are allowed) to do at various levels. James Paradis, David Dobrin, and Richard Miller describe the differences in writing that this hierarchical situation produced in an engineering facility they studied:

> it was the staff engineer's job to produce high-quality data; it was his or her supervisor's job to show how the group's activity was part of a coherent program; and it was the manager's job to make sure that this cycle of knowledge production was meeting corporate needs and making the best use of available resources. (1985, 286)

Paradis, Dobrin, and Miller's observation makes clear that different levels of the engineering organization are responsible for generating different kinds of capital, a phenomenon I demonstrate in this book. The organizational hierarchy, a part of the context in which engineers work, positions them to do certain kinds of intellectual work and writing. Indeed, the hierarchical organization may be seen as a reorganized social world that enables and shapes the nature of high-level technical work. As Hutchins (1993, 1996) suggests in his work on distributed cognition, social structures are also knowledge structures.

As shown in chapter 2, the hierarchy in effect at Pacific Equipment had an impact on its intellectual work, despite the fact that the company attempted to be less hierarchical than many large manufacturing companies. The company's effort was reflected, for instance, in its use of the term *team leader* rather than *supervisor*. However, hierarchy still shaped the work, and specifically the textual work, of employees at various levels. That is, the further up the hierarchy one was, the larger the portion of one's work that was likely to be accomplished on paper because, as Paradis, Dobrin, and Miller (1985) demonstrate, at higher levels, one processes the texts that have come from subordinates in addition to preparing texts that will go to superiors. Moreover, at higher levels, one was likely to issue texts meant to shape the work of subordinates as well as to report one's own to managers.

The lopsided distribution of textual work is reflected in the offices of Paul's group. The technician's office (shown in figure 3.7) contained the most electronic equipment, the software engineer's office (figure 3.8) contained a moderate amount, and Paul's office (figure 3.9) contained the least. Rather than work with objects, he worked with texts. He was the most likely of the four to work in his office, the locale for writing. He spent, he said, about 80 percent of his time there. In contrast, while I observed him, the cubicles around him were often empty. He said that the software engineer and technician were "on a vehicle" someplace on the grounds. Their work was more likely than Paul's to involve hands-on contact with the actual object, whereas his was more likely to involve writing. Paul told me that it was typical of the group leaders in the electrical engineering department to maintain the "documentation," the notebooks in which complete descriptions of the objects were kept. In these notebooks, Paul pulled together and coordinated the work he had previously parceled out to several subordinates, thus using writing as a way to manage both the intellectual and organizational hierarchy in his department. Additionally, Paul saw his role of maintaining the documentation as functional because it enabled the software engineers to focus on writing software.

Engineers' textual management and appropriation of technicians' knowledge was also reflected in Dan's work. Although Dan said that "even a good mechanic" could hear the same problems with a test that he did, Dan was valued as an engineer because he could also manipulate the paper forms of the vehicle and design the test protocol by which those paper forms were generated. Indeed, part of what Dan used writing to do was to manipulate the lab technicians who would, in turn, manipulate the object. Dan did this by writing work orders that laid out the tests to be run. He said that he had found through experience that these were best written in what he called "a cookbook style. . . . That's what works best because then

Fig. 3.7. *Office Belonging to a Technician*
in the Electrical Engineering Department

Fig. 3.8. *Office Belonging to a Software Engineer*

Fig. 3.9. Paul's Office

they don't have to read between the lines. If they have to read between the lines, you get this [pointing to a paper tacked up in his office that says 'I know you believe what you think I said. But I am not sure you realize that what you heard is not what I meant']." For this work, the lab technicians were assumed to do only a limited kind of thinking; the work orders were meant to make the technicians into extensions of Dan's hands, although his hands actually touched only the paper on which he wrote. For instance, as I observed him writing a work order, he said, "I need to remove some teeth from a gear," meaning that he had to get someone else to remove the teeth. He did so via a written work order and marks he made on an engineering drawing. When these work orders were executed, Dan's actions would be carried out by a lab technician. I will examine the work of the lab technicians more closely in chapter 4, where I will show that their work actually contained much independent effort. However, it is arguably true that Dan designed, initiated, and analyzed the work the technicians did. Writing and hierarchy thus multiplied the value of Dan's engineering training to Pacific Equipment.

Greg's coworkers talked specifically about lab technicians as extensions of their own work. The meeting at which I observed them took place just before the facility was to be largely shut down for a two-week hiatus. They were concerned that tests they needed run would come to a halt because no X13 would be available. I puzzled over this term until the context made it clear that an X13 was a lab technician. The engineers at the meeting evidently found the term somewhat opaque, too. One said that it sounded like "a device to measure with," and another joked that that's what it was. This joke seems to have an epistemological point to it. The technicians' status as semidevices not only extended the working reach of the engineers; it also protected the purity of the data they produced. Like (other) instruments, technicians theoretically had no stake in the outcome of the tests they performed but only had to perform them well. Thus, having a limited knowledge of the work they performed was assumed to be functional in the technicians' jobs and in the intellectual hierarchy into which they fit.

I must emphasize that Greg's coworkers did not see the work of the lab technicians as beneath them. Indeed, the source of their frustration in this situation was that although lab technicians would not be available to run their tests, union-negotiated work rules would not allow the engineers to run the tests themselves—something they believed themselves capable of doing and expressed willingness to do. To some degree, the hierarchical organizational context in which they worked encouraged them to treat the lab technicians as instruments: The organization generated positions of power and arrayed individuals within them. Interestingly enough, some of the conversations I heard made it clear that lab technicians did not always docilely function as extensions of the engineers' hands. Dan spoke of one technician as particularly difficult to work with (although he did not specify the nature of the difficulty) and spent time on the phone with the technician's supervisor planning how to get the work done as he wanted. Although the technician's resistance might challenge the organizational hierarchy at Pacific Equipment, it never seemed to disturb the intellectual hierarchy, which appeared more robust. Lab technicians could subvert the technical work that engineers had set out for them to do but not the paperwork that the engineers themselves did, because reports were not usually distributed to the lab technicians in either draft or final form. The only interpretation of the technicians' work to which the company granted legitimacy was the one performed by the engineers. Bourdieu argues that a speaker or writer has to be in a socially authorized position to be allowed to speak in certain ways (1991, 69, 72). Within the company, the technicians were not positioned to be heard as interpreters of data. Thus, it was difficult for the technicians to challenge the interpretations of objects that the engineers reached in their paper world. Indeed, I saw no sign that technicians wanted control over the interpretation of data, which they viewed as engineering work. As I will show in chapter 4, they wanted control over their own work, not that of the engineers.

While technicians could not challenge engineers' interpretation of data, these interpretations could be challenged by managers who, as I said in chapter 2, required documentation in the form of reports from engineers, just as engineers issued work orders to obtain data from technicians. As engineers interpreted the technicians' work for their purposes, managers interpreted the engineers' work for theirs. The engineers were clearly aware of this process and tried to meet the managers' needs while maintaining control over their own work, which they valued and largely enjoyed. Thus, Dan said that part of his purpose in writing to managers was to keep them both informed about what he was doing and satisfied that it was the right thing to do. For example, in a report on a persistent problem with a gear, Dan said that he included both the facts and a plan of action so that "management doesn't come down and say 'what is going on?' I don't want them telling me how to do my job." Dan's writing was intended to convince them that his place in the intellectual hierarchy was being well filled—that he "know[s] what [he is] doing."

Similarly, when Paul wrote an E-mail message to managers, saying that a device upon which his group was working had passed certain tests, he included a fair amount of detail about how the tests were run. That information, he told me "is meant to instill confidence in the reader that I was sufficiently deliberate in my investigation. I did the appropriate checks, everything I was supposed to do. So I'm persuading them that I did at least what they expected me to do." Paul has to persuade managers of his competence and diligence if he wishes to maintain control over both his actions and the interpretations of data he creates.

Writing both up and down the corporate hierarchy, then, engineers used paper forms of the object to try to maintain control of its artifactual form and to build into it the "quality" that they cared about. Responding to work orders from the engineers, laboratory technicians produced instrument traces or other test results. These pieces of writing then were interpreted by engineers in a joint process that was taken to reveal the knowledge believed to be embedded in the instrument traces. Thus they amassed knowledge of the object by using texts to appropriate the work of the technicians. This knowledge was subsequently shaped into reports that would convince managers that engineers' knowledge was valid and should be left in their control.

The hierarchical structure of this engineering organization thus affected engineers' writing. Moreover, it was to some degree made possible by writing. Latour points out that inscriptions can be amassed to intensify the power that their owner holds. Indeed, he sees that as their chief function (1990, 56–57). At Pacific Equipment, the process Latour describes is evident. Instrument traces allow engineers to gain power over objects and over the labor of the laboratory technicians because they amass the results of that labor in engineers' hands and serve as the basis for their interpretations. Reports by engineers to managers then allow the latter to gain power over the work of engineers and technicians, although engineers try to word their reports to maintain control over engineering knowledge.

CREATING KNOWLEDGE WITHIN AND ACROSS FIELDS

Pacific Equipment engineers, then, tended to treat texts such as instrument traces and engineering drawings as substitutes for artifacts that simply replicated the object so that the engineer could work on it more easily. Doing so put powerful arguments within the hands of the engineers. Data in the form of instrument traces are a highly valuable form of cultural capital that engineers have been educated to generate and then to treat as the solid foundation of knowledge. The engineers tended to believe that reasonable people would all see data in the same way if only they were perceptive enough. They saw their day-to-day interactions with coworkers as exchanges of factual information in which persuasion was not necessary and the exercise of power was not even conceivable.

Pacific Equipment engineers believed that they had to try to affect other's knowledge and actions only when an audience was likely to resist seeing the object in the way the engineer had come to see it. Thus they did not think that their control over the technicians required any particular kind of effort because they believed that hierarchy made technicians unable to challenge their knowledge, and thus they did not see themselves as needing to persuade technicians of anything. Rather they believed they could control the technicians' actions through instructions. I would argue that chapter 2 shows that managers used a fair amount of persuasion in shaping the work of the engineers; but engineers saw the technicians' work as manageable through instructions. When I asked an engineer how he would account for this difference, he speculated that it was partly owing to the fact that almost all of the managers at the engineering center were once engineers and thus feel some continuity with the engineers. In contrast, the engineers were never technicians, and a class disjunction appears to separate the labs from the engineers' offices. This engineer further speculated that managers believed that engineers had to understand the actions they were being asked to undertake, whereas engineers believed that technicians simply carried out orders.

Moreover, unless there was a problem, efforts to persuade other engineers or to exercise symbolic power over them also remained invisible because agreement had been reached in the smooth interaction of everyday practice. Problems arose when things broke down or at the boundary between different kinds of expertise, where an expert in one area was likely to have trouble understanding an expert in another. The moment when texts moved from the field of engineering into the encompassing field of the organization was almost always one such occasion. Thus, these engineers believed that they needed to persuade management of the validity of their work because managers had a different kind of knowledge and concerns and because the organizational hierarchy meant that management was able to resist their conclusions in a way that others were not.

One sign of the way in which engineers saw their own knowledge generation as arhetorical was that when I asked several engineers about persuasion, they all saw data as uninvolved in any rhetorical process. As I mentioned in this chapter's opening, data are the coin of the realm in engineering. They are a form of cultural capital that engineers tend to see as of obvious value in itself rather than as a persuasive tool. Preserving the "pure" nature of data is important to engineers because that purity validates their own form of knowledge. They all knew that they sometimes used data from instrument traces as evidence in preparing other documents; however, they did not believe that doing so made the data persuasive. For instance, in vignette 3, Derek says that engineers will need more data about their project before they can expect upper-level manager Ken to give them a budget. However, engineers were unlikely to say that the need to persuade Ken made the data persuasive. Rather, they divided the evidence from the goal at which it was aimed and said that such writing was descriptive writing aimed at persuasive purposes. Paul, Greg, and Dan all theorized about their use of persuasion in this same way. Thus, in explaining why he included details in a test report, Paul said, "The information itself is descriptive, but the fact that it's there is persuasive." He used similar terms in accounting for another engineer's report on a software problem: "The main thrust would be descriptive, but the very fact that he has to write that to another engineer is persuasive to some degree."

Greg, too, distinguished between a document's descriptive content and its persuasive purpose. "I would say in meeting notes, I usually don't write any persuasive notes," he told me. "I do write descriptions of how I want to persuade somebody of something though." His comments on the use to which the real-time pressure distributions were put were especially interesting. Recall that these were the colored pictures he and his coworkers discussed at a meeting. These visuals were later used in a meeting between Greg's group and representatives of an outside company that supplied them with the parts that were represented in the pictures. "I would like to call those descriptive," Greg said, "but they were used in a persuasive manner. . . . And in fact, the more times they were used, the [more] persuasive they became." By this, Greg meant that he and his coworkers first negotiated with each other to decide what the pressure distribution pictures "showed." Next, they used them to "gain support within" Pacific Equipment for their understanding of the problem. Finally, they used them to change the supplier's idea of what the problem was. Greg saw these uses of the pressure pictures as progressively more persuasive as he and his coworkers became increasingly certain about what they meant. Increasing certainty was both created and signified by the growing network of relevant people who accepted a single interpretation (cf. Latour 1987). Joint interpretation of instrument traces was the means by which he and his colleagues created something they were willing to call "knowledge." As was typical at Pacific Equipment, such creation of knowledge was made easier by the common disciplinary understanding that the engineers shared.

Like Paul and Greg, Dan, too, distinguished the evidence in a document from the persuasive purpose to which such evidence might be put. In talking of a report he had written to management, for instance, we had the following exchange:

> Dan: I'd say that was pretty much descriptive. Persuasiveness is somewhat hidden in that when this gets to management . . . [it's meant to convey] that we don't have the answer to this problem and we'd better get some time and bodies to work on it. . . . It's also providing them enough description to convince them . . . you know what you're doing.
> Dorothy: So you use the description for a persuasive purpose?
> Dan: Yes.

The engineers I observed tended to see persuasion at work only when their audience was likely to resist whatever claim they were making and not when their audience was likely to agree with them. Thus they did not see persuasion at work when coworkers negotiated agreement about what visual evidence meant. Such events as Dan and Tom jointly analyzing a strip chart or Greg and his coworkers doing the same with the real-time pressure pictures were transparent for them. They tended to see knowledge in those situations as coming straight from the object as represented by the instrument traces. The engineers simply had to look at the traces perceptively enough to see what was there. So, for instance, when I asked Paul if a report that one of his subordinates was sending to an engineer in another group had to be persuasive, he said: "Once he checks into it, we are confident that he will find the same thing [we did], but we are trying to persuade him that there is a problem. We don't have to try very hard, and we don't feel that persuasion is very necessary." Thus, engineers valued these written documents but saw nothing rhetorical about them or about the actions surrounding them although the acceptance of these texts as self-evident meant that they could actually be used as extremely powerful pieces of evidence. Within engineering, coworkers shared enough common values and understanding that persuasion seemed unnecessary. Of course, such a conclusion ignores the large amount of social effort that went into producing those common values and understandings, effort that becomes quite visible, for instance, when we look at the work of summer interns, as I do in chapter 5.

These engineers were most likely to see persuasion at work when they wrote to management, the part of the organization that could and occasionally did resist their claims to knowledge. Thus, Dan talked about test reports written to management in the following terms:

> Dan: You've got your conclusion. And then you have to back it up with descriptive information to support that. Because you're trying to persuade management based on that conclusion.
> Dorothy: So you would say that most of your persuasion, most of the time when you're being persuasive, it's directed at management rather than at other engineers?
> Dan: Yes. The other engineers, you include them in the mailing list because it's information that may help them in the future.
> Dorothy: But they're not necessarily who you're aiming at in the report?
> Dan: Right.

These engineers believed that persuasion was most likely to be necessary when they wrote to managers with their different sense of priorities and their organizationally granted positions of power. In contrast, within engineering, writing that was closely linked to the object was perceived as a substitute form of the object and not perceived as persuasive, although, for engineers, evidence that has been positioned as unmediated is the most powerful symbolic exchange possible.

POWER, KNOWLEDGE, AND ENGINEERING GENRES

In the observations I have described here, writing tends to fall into three divisions or what we might call "three families of genres," one of them privileged within the field of engineering, one of them used to bridge the gap into the field of the larger corporation, and one of them used to draw upon the actions of laboratory technicians. Within engineering, the most valued texts are fragmentary forms such as instrument traces that come from the lab and create the illusion that reality has written itself down for engineers to analyze. For these engineers, orally negotiating consensus while actually looking at instrument traces appears to play a large role in defining knowledge. If they and their colleagues could reach joint interpretation of the instrument trace in front of them, then they were able to claim a tentative kind of knowledge, although their knowledge eventually needed to be validated by the performance of the object they designed and released into the marketplace. Data gathered through instrumentation represented the kind of almost certain knowledge that engineers valued within their own field. Engineers have a disciplinary commitment to achieving certain knowledge, even though their daily work teaches them that such certainty is always elusive and temporary. The object can always malfunction; it can always be improved. The engineering ideal, however, is arhetorical, and it is the goal of engineering to make knowledge as certain and undebatable as is humanly possible.

Crossing the gap into the larger corporation, however, engineers valued extended pieces of prose, such as reports, in order to prove their competency. Such reports were an institutionalized requirement from managers, but engineers attempted to shape them for their own ends. In these reports, engineers generally aimed to influence the way in which managers structured technical work, although managers, especially those who were more than one level removed from the engineers, were often not expected to understand the intricate details of that work. Such expertise was, after all, what the engineer was hired to provide. If engineers had reached consensus within the engineering department, managers usually accepted the technical conclusions they had reached, while reserving the right to make their own decisions about program direction. For engineers, the challenge was to create engineering knowledge while simultaneously striving to maintain its value within the for-profit corporation. Thus in vignette 3, John and his colleagues have to use data both to build engineering knowledge about their vehicle and to convince managers to pay for whatever redesign they decide is necessary. The second goal is often more difficult to reach than the first.

Thus, engineers at Pacific Equipment rely on two different families of genres in establishing their knowledge as valid, one kind addressed to coworkers and the other addressed to the corporate hierarchy. They share common values and daily experiences with coworkers that help them to form similar judgments about what actions are desirable. However, they find that managers often operate from different assumptions. Thus they write in at least two different basic social situations. These two different sets of social situations are met with two different families of genres: data reports for oral interpretation with their coworkers and more sustained prose in reports justifying their actions to managers. In a different study, I found that these two kinds of genres are also those that novice engineers wrote most often (Winsor 1996, 27–38), suggesting that engineering contexts might commonly make use of this division of textual labor.

Additionally, the engineers use a third genre, work orders, in communicating with laboratory technicians. I will discuss this genre further in chapter 4, but it should be clear from what I have said in this one that engineers believed that work orders should function simply as instructions and that any failure in this process could be attributed simply to unclear instructions or to technicians' inept reading. Vignette 4 will show a technician drawing on a far wider range of skills than such a view allows.

For Pacific Equipment engineers, a sense of what is rhetorically necessary varies depending on their positioning in the organizational hierarchy relative to their readers. When they write to those above them in the corporate hierarchy, they must work harder in order to generate some control and power for themselves because this social capital is not provided by their role in the organization.

The social contexts in which any of us work shape our understanding of how we build knowledge and shape our generic texts. For engineers, these contexts include both the field of engineering itself and the larger corporate field within which engineering is usually nested. Similar situations probably exist for many professionals working within corporations. Technical writers, for instance, must frequently struggle to convince managers that funding usability testing represents a worthwhile exchange of monetary capital for the cultural capital of knowledge. At Pacific Equipment, engineers use characteristic texts within engineering and within the larger corporate field to build knowledge and to convert the capital that comes with it into symbolic power.

Vignette 4

Two Hours in a Technician's Afternoon: Rich

I arrive at the engineering center a few minutes before 12:30 and check in at the security desk. The security guard phones Rich, the technician I will be observing today, and he comes and gets me and escorts me back to his area. Rich maintains the inventory of parts that lab technicians can check out for use in their work. We enter a large room with rows of shelves full of parts. Rich has a desk there with a computer and shelves with catalogs from which he can order parts. Off of this large room is a smaller one where we will spend most of our time today. Here, Rich builds parts that are needed to conduct tests. These parts are used in the test cells, in vehicles, and in the engineering area. Today, I will watch him build several parts based on engineers' work orders.

I laugh as we enter this smaller room because the first thing I see is a voodoo doll with Rich's name on it hanging from a hook in the ceiling. There are several pins in the doll. In the center of this smaller room, there is a long worktable. There is also a basket of drawings. Rich takes one out to use with a work order he is working on. The engineer has sent minimal instructions (one line saying basically "do the same thing again") because he knows that Rich has built this part before and already has the drawing. It turns out that there are several versions of this part, so Rich calls the engineer to make sure he has the right drawing.

Rich says that some engineers always send accurate work orders and others are less reliable. He shows me some of the work orders. Some have elaborate, detailed computer drawings. Others are little hand sketches. The one that he is working on is rather simple. He's making two copies of a cable. As he works, another technician comes into the room and asks if Rich knows where a math conversion table is. He needs gallons/minute and has cubic centimeters/engine revolution. Rich doesn't have the table. The other technician leaves to reappear in a few minutes saying he thinks he has the conversion but he needs a flow meter. Rich tells him where they're stored in the next room but the other technician says Rich doesn't have the right size. They confer on how to improvise a solution to deal with the problem.

As they finish, the other technician eyes me curiously and then asks Rich if he has a "new helper." Rich goes back to work on his cable and simply answers "yes." The other technician waits for a second or two but no more information is forthcoming so he leaves. I ask Rich if he is asked to explain my presence after I leave. He laughs and says that people are usually too polite to ask who I am while I'm there.

Rich finishes his cable and then phones the engineer who ordered it, leaving voice mail that the work is done. He starts another job, which is interrupted several times by phone calls, mostly about the inventory he maintains. The engineer for whom he created the first cable enters the room to pick up his cable and happily leaves with it.

Rich studies the drawing he is currently working from, and then opens a drawer full of parts, intending to choose the right one. "Something as simple as putting a terminal on a wire," he says, "you would think 'How tough can that be?' But it's tougher than you think because of the details. You might need a specific tool to crimp the wire, for instance." He talks about the training that technicians undergo. A class is currently being taught by one of the technicians in the use of a new scope. Rich finishes his current task and again phones the engineer who requested it. "What am I gonna do?" he asks. "I'm out of the simple jobs."

He takes out another work order. He phones the engineer who wrote it to ask for a special part that he needs. The engineer says he'll bring it at once and is there within five minutes. While he is there, Rich shows him a second order he wrote and asks for clarification. They look at the order together. The engineer wants the part built as it was completed by another technician in the past. Rich asks about size; the engineer says that the past part could be held in the hand. Rich asks what kind of switch the engineer wants. The engineer mimes with his hand to test for what switch would be best. They discuss the availability of different kinds of switches. After the engineer leaves, he says, "You have to ask this guy questions. . . . He can always answer your questions, but the knowledge doesn't always make it to paper."

Before Rich starts building this part, he takes advantage of this break in his work to walk me out of the building. It's 2:30.

CHAPTER 4

AMASSING KNOWLEDGE IN THE HANDS OF THE MORE POWERFUL

There's two different worlds, the engineers and us. You're always
trying to bridge that gap. "You wrote this. What does it mean?"
—Pacific Equipment Technician

interpretation

What makes power hold good, what makes it accepted, is simply
the fact that it doesn't only weigh on us as a force that says no,
but that it traverses and produces things, it induces pleasure,
forms knowledge, produces discourse. It needs to be considered
as a productive network which runs through the whole social
body, much more than as a negative instance whose function is
repression.

—Michel Foucault, *Power/Knowledge*

These opening words were spoken by a Pacific Equipment lab
technician. They raise questions about how social and textual
structures are intertwined when people write to one another within
complex, hierarchical organizations, a question which, as I indicate
in chapter 1, rhetoricians have answered in terms of genre studies.
In this chapter, I further examine the question posed by Aviva
Freedman and Peter Medway that I cite in chapter 1: "How do some
genres come to be valorized? In whose interest is such valorization?"
(1994, 11) Chapter 2 shows that people use genres to control the
capital that is important to them. Chapter 3 shows that genre
choice varies depending on how writers are positioned in relation to
readers. In this chapter, I argue that, in an organization, the texts
that different people produce are not equally likely to be regarded by
the organization as genres. Carol Berkenkotter and Thomas Huckin

argue that genre is not an all or nothing category, but rather is a continuum along which more or less typified texts can be ranged (1995, 17). But placement on the continuum is not a straightforward matter. Whether a textual form is perceived as a genre is more than a matter of how common its use is. In addition, that common use has to be *visible* to a significant number of people in an organization. In effect, I will argue that organizations convey generic status on texts as part of the way in which they create positions of power. As John B. Thompson argues, fields assign value to linguistic products and the most valuable are the most unevenly distributed (1991, 18). In her groundbreaking article on genre, Carolyn R. Miller pointed out that whether a situation and responding text are perceived as recurring and typified is intersubjectively defined by the people involved (1984, 156), but genre theory has not explored this insight in any systematic way. This chapter argues that among the factors affecting recognition of genre are how important a social system perceives the text's function to be and (not unrelated) how visible its users are.

The visibility of particular texts and writers is part of the order that is constantly renegotiated in any social system. That is, visibility (or invisibility) is a state that is constantly reachieved by both one's own actions and those of others. As Anthony Giddens pointed out, human agency is at work in the organization of society, but participants do not start from scratch. Rather, they operate within already existing situations and "reproduce or transform them, remaking what is already made in the continuity of *praxis*" (1984, 171). In so doing, they shape the order in which they function, a process that among other things generates relations of power. Thus even trivial, everyday actions can have profound effects on the social surroundings that people inhabit.

In chapters 2 and 3, I talked primarily about engineers and managers at Pacific Equipment but I also mentioned laboratory technicians who collected the data that the engineers analyzed. In a study of the laboratory technicians who worked with Robert Boyle, the historian Steven Shapin (1989) pointed out that the work of people like technicians is often invisible to us, that it is subsumed in the work of the supervising scientist, engineer, or manager. Thus we say that Boyle conducted certain experiments or operated his vacuum pump when what we really mean is that he ordered

Visibility

Separation

others to accomplish these actions. Chandra Mukerji (1996) offered a more contemporary example of the same phenomenon when he examined the way in which the chief scientist in an oceanography lab was constituted as a scientific "genius" when the collective work of the lab was attributed to him.

As I will demonstrate in this chapter, at Pacific Equipment, work done by blue-collar workers in the lab was similarly folded into the work of the engineering area. There was tension between these two regions of the engineering center because, while their work needed to be oriented toward the same goal (the development of improved off-highway equipment), their interests were not identical and they were not always seen as having equivalent agency in the work they did. Indeed, as I suggest in chapter 3, the engineers sometimes saw the technicians as tools they activated through work orders, rather than as agents in their own right. What's more, they valued this tool-like status for the technicians because it contributed to what they saw as the objective nature of the data technicians took. The engineers and the technicians were hierarchically divided, with the technicians working for the engineering area but not being a fully fledged part of it. In effect, the technicians stood in the same relation with the engineers that the engineers did with the managers. Like the engineers, the technicians drew on a considerable body of specialized knowledge and sought control over their own work. However, the organization positioned the engineers to direct the technicians' work and, in some ways, to claim credit for it. This chapter explores the way in which a particular genre, the work order, was used to make use of the technicians' skills while maintaining the existing social structure at the engineering center.

The genre of the work order was a discursive tool that simultaneously allowed the technicians' work to be done despite any differences they might have with engineers and that maintained the hierarchical structure of the engineering center because it both triggered and concealed the work the technicians did. To echo Giddens (1984), work orders were structured discourse (i.e., they were generic) that helped to structure the organization. They served as an ordering tool for the relationship between the engineers and the technicians, mediating their relationship and serving as a concrete representation of their interaction, a function that Charles Bazerman has identified as one of the uses for texts in organizations:

Mediation

Because the produced discursive objects are in a sense
concrete, although symbolic . . . they provide a concrete
locus for the enactment of social structure. That is, whatever individuals may feel and think about each other,
however they may sense they relate to each other, whatever beliefs they hold about social hierarchy and obliga-
tions, however they may perceive social pressure and
power, there is still an observable, recordable, collectable
utterance that concretely mediates among these various
personal orientations. (1997, 297)

In this chapter, I will first give some details of how I conducted
this part of my study. Then I will talk about the nature of the technicians' work in a hierarchical organization; the work order as a
generic textual tool within this hierarchy; and the contrast between
work orders and other texts that the technicians themselves produced, texts that were treated less as typified responses to typified
situations (i.e., as genres) than as individual, idiosyncratic creations.
The chapter will show how one organizational genre, the work
order, functioned to orient technicians and engineers so that joint
work could be accomplished despite the tensions between them,
and so that the joint work could be credited within Pacific Equipment Corporation's normal hierarchy. Like chapters 2 and 3, this
one emphasizes the political aspect of genre as a form of social
action, an aspect our research and theory have tended to neglect.

STUDYING WORK ORDERS

In the Pacific Equipment engineering center, the labor of generating data was divided between the engineers and the technicians,
with the work of the former conceived of as mental and the work of
the latter conceived of as hands-on. As I showed in chapter 3, engineers wrote work orders that the technicians carried out. In vignette
4, Rich is using work orders. The work orders were necessary not
only because of the division of labor that was enacted in Pacific
Equipment but also because of the physical layout of the facility. At
the company, engineers and technicians were located in widely separated areas of the large engineering center, creating what Giddens

calls a "regionalization" of the slice of time-space they mutually occupied (1984, xxvii; for other discussions of the impact of spatial organization on discourse, see Chin 1994, or Cross 1994). Such an organization of space is not inevitable. Indeed, at one meeting I attended, engineers were talking enthusiastically about a new engineering center being built in another state. The center was being designed so that vehicles could be moved into the engineering area for hands-on examination. And one engineer, Dan, told me that he sometimes thought he should move his desk out into the labs because then he could be in closer touch with what was happening there. However, the design of the Pacific Equipment engineering center reflected the notion that mental and hands-on work were separable, and indeed that the latter was subordinate to the former.

In accord with this idea, the engineering center's numerous labs and mechanical areas were widely separated from the engineers' cubicles. In the labs, technicians performed a wide variety of tasks including building experimental parts and testing them. Testing was done in specialized "test cells," that is, rooms holding equipment that measured and recorded the performance of Pacific Equipment Corporation products under development. This equipment ranged from large refrigeration units to test how well the vehicles ran in cold weather to dynamometers, which were the equivalent of treadmills for equipment such as engines and transmissions. On the dynamometers, the products could be run at various speeds and conditions to test their durability or power, for instance.

I observed three Pacific Equipment technicians in the summer of 1997:

- Gary, who operated equipment in a test cell;
- Jim, whom we see in vignette 1, and who built and modified engines; and
- Rich, whom we see in vignette 4, and who built custom instruments for the test cells and maintained a supply center to which other technicians came to get parts they needed for their work.

In order to recruit these participants, I had to proceed somewhat differently than I usually did. All my other participants volunteered for this study in response to E-mail distributed by the Human Resources

(HR) Department. However, I had to recruit the technicians in a more mediated fashion. The HR manager first contacted the technicians' supervisor. This supervisor then recruited the participants. One basis for their selection was that Gary and Jim were union shop stewards and Rich had been a steward at one time. The supervisor evidently felt that if these union officials were willing to cooperate with me, I would be better received in the lab area. Such a consideration shows the supervisor's awareness of the possibility for contradictions between the goals of management and the technicians, as the existence of the union signifies the technicians' awareness of the same possibility.

During my observations, I read the multiple work orders that the engineers wrote and that were on the technicians' worktables. When I was working with the technicians, they read through some of these with me and offered interpretations and evaluations. Two months after I finished observing the laboratory technicians, I was also able to interview the engineer-manager who was responsible for coordinating work between the lab and engine engineering, along with an engineer who worked for him and was in the process of examining engineering/laboratory relations, including those generated by work orders. The coordinating engineer provided me with sample work orders that he cleared for publication. Additionally, other technicians who knew about my project approached me and talked to me about work orders when they saw me at the engineering center. Work orders were also occasionally topics for discussion during other summers I spent as an observer in the engineering center. Both engineers and technicians believed that work orders were important at Pacific Equipment and hoped that my research would offer some insight for improving them.

THE WORK OF TECHNICIANS AND THE HIERARCHICAL DIVISION OF LABOR

As an engineering design center, this Pacific Equipment facility valued theoretical, symbolic knowledge. As one engineer said, "We make two things at [the engineering center]: We make the drawing and we make the spec[ification]." In other words, the engineering center does not make Pacific Equipment products itself; that function is performed by the company's assembly plants. Instead, the engineering center creates knowledge and symbolic representations of that knowledge. This abstract goal and the hierarchical division of labor through which it is accomplished made it hard to keep the technicians' more hands-on work in mind, an omission that was functional in maintaining the corporate hierarchy.

In Pacific Equipment , as in many technical workplaces, a portion of the work is carried out by technicians, skilled workers whose hands-on work is seen as supplementing the mental work of white-collar scientists, engineers, and managers. As is common in such workplaces, the company conceived of hands-on work and mental work as separate and hierarchical steps toward its goal (see Marvin 1988, 9–15, for a discussion of the cultural history of this hierarchy). The technicians' task is to turn material reality into data. In addition, technicians are also often responsible for taking care of the material entity itself and making sure that it is working correctly (Whalley and Barley 1997, 47–48). Because their part of the work is initiated and used by the engineers, it can be difficult to keep sight of their contribution. The data they generate are seen as belonging to the engineer, not the technician, or, as Stephen R. Barley and Julian E. Orr put it, the technician's job is to "reduce material phenomena to information that becomes grist for the mill of another occupation" (1997, 14).

While engineers and technicians got along well at Pacific Equipment, both groups spoke of occasional problems that seemed to stem from technicians' resistance to organizationally established relations of power. In vignette 2, for instance, Doug, an upper level manager, accepts the fact that lab workers may show a certain amount of hostility toward managers. On another occasion, an engineer told me that his audience in the lab were people "who can't read or refuse to read instructions" (cf. Hull 1999 for a similar case of an engineer interpreting failure to follow instructions as a sign of deficit in blue-collar workers). He also said that one technician consistently "bucked the system" and that the department in which this technician worked "prides itself on the number of supervisors they go through."

If the engineers sometimes found the technicians resistant to their authority, the technicians sometimes found the engineers arrogant. One of the technicians, Rich, said that newly hired engineers often did not know what they could expect from the technicians. He said, for instance, that they might not know there was a union shop and that there were work rules saying what the engineers could and could not do. He said that there had been a plan for new engineers to work in the lab for a few hours in order to get a sense of what technicians' work was like, but that this plan had been resisted by the technicians who were afraid the company might be planning to have the engineers do their jobs if there were a labor dispute. All of this suggests that the technicians did not always accept the relative positions of engineers and technicians in the Pacific Equipment hierarchy. Thus, various company practices, including textual practices, had to be established to reinforce their relative statuses if they were to remain effective.

The statuses of the engineers and technicians at Pacific Equipment were marked and thus enforced in a variety of ways.[1] Engineers worked in carpeted cubicles, for instance, while technicians worked in uncarpeted areas with cement floors. This particular contrast also had a practical reason, of course, in that carpeting would not have survived for long in the lab environment. Moreover, the engineers' workspaces were hardly luxurious; a carpeted cubicle is still a cubicle, lacking the walls and door that marked the spaces of their managers. More significant perhaps were the facts that the technicians' time was more regulated than that of the engineers (e.g., they took breaks whose beginning and ending were signaled by a buzzer), and that while engineers frequently entered the lab, technicians seldom went into the engineers' area. Both of these facts meant that the technicians' work was scrutinized not only by their supervisors but also by the engineers. In Foucault's (1979) terms, the engineers could exercise a governing gaze while the technicians could not. At least one of the technicians explicitly connected the visual observation of someone else's work with its regulation. He told me that if he were as rich as Bill Gates, he might "hang around and annoy" his boss by sitting in the cafeteria and "glaring at him when the break buzzer rings." In contrast to the engineers' scrutiny of their work, the technicians commented on the engineers' work only to one another (and to me). This lack of officially recognized upward scrutiny echoes the situation I noted in chapter 2, in which managers' actions sometimes remained undocumented so that they would be less available for scrutiny by engineers.

The technicians' primary function was to generate data for the use of the engineers (Barley and Orr 1997). For instance, a technician might run tests to determine if a new piston the engineer had designed would increase the amount of power the vehicle provided. He or she would measure the power output of the vehicle at various engine settings the engineer had chosen and then communicate these data to the engineer by means of a written test report. At Pacific Equipment, test reports were almost always lists of numbers. In some test cells, technicians recorded data by hand; in most, they generated the lists of numbers by using the computer to take data at points the technicians preset. According to Gary, one of the technicians I observed, technicians had some choice on whether to use the computer, and some refused to do so because they believed that using the computer would mean they were less "in control" of their own work (cf. Zuboff 1988, 63). As I suggest in chapter 3, control over one's own work is important to most of the people in the engineering center.

While computer-generated data made some technicians feel they were less in control, such test reports reflect a creation of meaningful text that is less automatic than it might appear. For instance, when Gary used the computer to set up the tests in his cell, he chose from a menu of possible tests and entered various parameters, some from the work order and some from his own knowledge. He not only had to set the computer to take the data, but he also had to be able to look at the data and tell if the various machines were working correctly. Thus the technicians participated in generating knowledge and meaning when they got the data onto the page. While the technicians generated these reports, however, the engineers did not see technicians as the "authors" of these reports. Rather, they saw the numbers as belonging to the engineer who had ordered the test and thus had used the work order to remotely control the technician. They saw these reports as authorless, or perhaps as being generated by the equipment in the test cell, a category into which they sometimes placed the technicians as well as the machinery the technicians used. For example, as I said in chapter 3, I heard a group of engineers joking that the term by which test operators were called, an *X13*, not only "sounded like a device to measure with" but *was* a device to measure with. (See Sullivan 1997, for a comparable account of Internal Revenue Service clerical workers whose instruction manual treated their work as unrealistically mechanistic.)

As we saw with reports that engineers wrote for managers, texts can be used to create hierarchy because they can create the "immutable mobile[s]" Bruno Latour describes (1987, 227). Like those engineers' reports, the data reports from the lab can be used to keep a representation of reality constant while moving it into the control of centers of organizational power. Indeed the ability to use such immutable mobiles is one of the things that creates and defines a center of power. Being in a powerful position may allow one to use the knowledge someone else has generated, but being able to use that knowledge is one of the things that generates the powerful position. Thus when a written report of technician-generated data is moved to the engineering offices, the engineers are able to use far more knowledge than they could have generated by themselves. As a tool, texts are thus particularly good at allowing the technicians' work to disappear into the engineers' work.

WORK ORDERS AS GENERIC TEXTUAL TOOLS

As a routine part of their work in the organizational hierarchy, then, technicians read and carried out work orders. These work orders constituted a genre; they were a typified textual response to a typified social situation. Their typification was indicated by the fact that they had a name. The name "work order" constituted a category into which varied texts could fit with almost no effort by anyone working at Pacific Equipment.

On the most obvious level, the purpose of the work orders was to establish the tasks of the technicians. Via work orders, engineers told the technicians to conduct a test in a certain way, for instance, or to replace a standard part on an engine with an experimental part. The one-way flow of work orders from engineers to technicians is an important but taken-for-granted tool that marks the technicians' work as subsidiary to that of the engineer. I saw dozens of such orders during the time I spent in the lab. Figures 4.1 and 4.2 show sample orders, with figure 4.1 showing an order for tests and figure 4.2 showing one for an engine rebuild. (I will point to these figures throughout the following discussion.)

As figures 4.1 and 4.2 show, test orders contain some repeated formal elements. At the top of the order, there was always orienting information—date, engineer's name, mail code, phone number, account number to which the test would be billed,[2] and start and finish dates for the testing. Although these same items always appeared in work orders, their order varied, as can be seen by comparing figures 4.1 and 4.2. This variability suggests that there was no company-wide template for work orders, although departments may have had one and individuals certainly did. According to Gary, most engineers saved previous work orders in their computers and used them as templates for generating new ones, a practice that was efficient but occasionally problematic because the engineer sometimes overlooked changes that needed to be made to adapt an old order for a new purpose. In addition to the orienting information at the top, elements in work orders also included subject lines and numbered lists. Gary saw the subject line's clarity and completeness as one of the factors making for a good work order, and the orders that I saw were usually arranged in numbered steps.

These common formal features, however, did not seem to be what marked these texts as a genre. Rather, their generic status flowed from the social action they repeatedly performed. They were always written by people in the engineering area to the technicians in order to set the tasks the technicians were supposed to accomplish. Everyone in the organization could recognize a work order from that social action even if the order were handwritten.

Work orders, then, were aimed at shaping the technicians' actions and, by means of those actions shaping the physical devices they worked on so that data could be gathered. While the written work orders initiated a situation in which certain work got done, however, they did not determine that work by themselves. Rather they were supplemented by a wide variety of material and social arrangements, including organizational and intellectual hierarchy, phone calls, engineers' visits to the laboratory, the craft knowledge of the technicians, and other forms of writing such as manuals, instruction sheets, marks on the machine parts, and photographs. Texts, such as the work order, interacted with other factors. However, work orders seemed to be the most visible representation of the process. Other factors were oriented around them and seemed subsidiary to them. Their very visibility and permanence were what

Fig. 4.1. *Work Order for Testing*

TEST LAB SHOP ORDER

Dept.___C___	Date ____14 Sept 1998____
Dept._____	Engineer _____
_____	Mail Code _____81J_____
Work Area ___2AX10_____	Phone _____
Operator _____	Account _____
Est. Start _____	Test Unit_____
Comp. Date_____	Parts Avail. ___14 Sept 1998___
Order ID _____	Due Date ____28 Sept 1998____
Shop Use Only	Initials _____
_____	Backup _____
	Backup Phone # _____

cc:

Subject: **XXXXXXXXXXX EMISSIONS CORRELATION TESTING**

This engine will be used for baseline testing in our emissions test cells. It was tested at XXXXX and will now be used to correlate emission test cells at XXXXXX.

THIS ENGINE MUST NOT BE CHANGED IN ANY WAY THAT COULD AFFECT PERFORMANCE OR EMISSIONS. DO NOT MAKE ANY CHANGES TO THE FOCUS ECU. ENSURE FOCUS ECU REMAINS WITH THE ENGINE.

1.   Instrument to measure:
    a.   Dyno load
    b.   Engine speed
    c.   Fuel weight
    d.   Fuel time
    e.   Fuel inlet temperature
    f.   Fuel outlet temperature
    g.   Air meter outlet temperature
    h.   Oil sump temperature
    i.   Coolant inlet temperature
    j.   Coolant outlet temperature
    k.   Compressor discharge temperature OR
    l.   Intake manifold temperature
    m.   Turbine inlet temperature
    n.   Turbine outlet temperature
    o.   Blowby
    p.   Smoke

q.	Oil pressure

q.   Oil pressure
r.   Air meter pressure drop
s.   Air meter outlet depression
t.   Intake restriction
u.   Compressor discharge pressure AND/OR
v.   Intake manifold pressure
w.   Exhaust manifold pressure
x.   Exhaust restriction
y.   Nozzle needle lift
z.   TDC, 30 BTDC, & 60 ATDC
aa. Engine hours—**Input 1,214 hours into computer at start of test.**

2. Use the ECU for all of this testing, droop 1 and torque curve A. Set dew point to 15° C for all tests.

3. Run a variable speed curve (2,200 to 800 rpm) in 200 rpm increments. Confirm correct engine operation with writer. The following table shows some expected values for full load rated speed:

Intercooler Outlet Temp: 65 ± 2 C (Need Automatic intercooler)
Intercooler Pressure Drop: 20 ± 2 Kpa

Speed: 2,200 rpm
Load: 867.7 ± 15 N-m
Power: 200 kW
Timing: 8.0 ± 0.5
BSFC: 204.4 ± 3 g/kW/hr
Fuel Rate: 41.1 ± 0.8 kg/hr

4. Run three 8-mode emission tests with particulates. (if there are control problems, run govenor in MIN/MAX). Run mode 8 at the free idle speed, i.e., 800 ± 5 rpm and no applied load on the engine. Peak torque speed is 1,400 rpm.

Use these operating conditions:
a. Air inlet temperature   25 ± 1 C
b. Intercooler outlet temperature, FLRS   65 ± 1 C
c. Coolant outlet temperature   92 ± 2 C
d. Fuel inlet temperature   40 ± 1 C
e. Air intake restriction   3 kPa @ full load, rated speed
f. Exhaust outlet restriction   7.5 kPa @ full load, rated speed

5. Run 3 (sets of 3) Federal Smoke tests using the Celesco smoke meter (ECU must be in ALL SPEED MODE). Record data using the strip chart. Set intake restriction to 6.2 kPa. Check with writer to determine whether additional FSC testing is required.

6. Do all other work required for completion of this order.

Fig. 4.2. *Work Order for Equipment Build*

TEST LAB SHOP ORDER

Dept: 029	Date: 05 May 1998
Order ID:	Engineer:
	Mail Code:

Work Area:	Phone # (Pager):
Operator:	Account #:
Act. Start Date:	Test Unit #:
Act. Comp. Date:	Parts Avail.: now
	Req. Start Date: 05 May 1998
Shop Use Only:	Req. Comp. Date: 06 May 1998

CC:

SUBJECT: Replace Damaged Piston/Liner and Repair Engine as Needed

Engine XXXXXXX was running in the dyno cell prior to failure. Please do the following to get the engine back in test condition:

1. Place the engine in a rollover stand.

2. Remove all engine components and test equipment that will hamper the removal of the cylinder head.

3. Remove all of the pistons and replace with the part number listed below.

4. Do any work necessary to complete this shop order.

Part Number	Qty.	Description
XXXXXX	4	Piston
XXXXXX	1	Piston
XXXXXX	5	Ring, Piston, Set
XXXXXX	1	Gasket, Cylinder Head
XXXXXX	2	Gasket, Exhaust Manifold
XXXXXX	2	Gasket, Exhaust Manifold
XXXXXX	3	Gasket, Intake Manifold

Approval:

gave power to the work orders. In this sense, they functioned like the immutable mobiles Latour describes (1987, 227).

THE PROCESS OF ISSUING WORK ORDERS

Work orders were supposed to be issued in a particular, institutionalized way. This changes periodically, but during the summer of 1997, when I observed the laboratory technicians at Pacific Equipment, the official story of how work orders should be issued went like this: The engineer wrote the order and sent two copies, one to the supervisor of the lab area in which the order would be carried out and one to the person who maintained the schedule of work in the lab. The supervisors and scheduler met once a week to prioritize the work and to allot it to the various test cells that were run by the laboratory technicians. As a technician finished one job, he or she was supposed to contact a supervisor who called the scheduler to get the next order on the list for the area in which the technician worked. Notice what this story of how work orders were issued implies: Because technicians were expected to take the next work order in the priority list, the official process implied that any technician in a given area could do any job, that they were interchangeable tools. In reality, this story of issuing work orders was a tidy fiction, a goal that people aimed to match but from which they departed when it interfered with their efforts to accomplish meaningful work. In this variation between official texts and unofficial action, the account of issuing work orders resembled a budgeting practice that I described in chapter 2. As may be recalled, the official budget called for an engine repair to be billed to the project for which tests were being run when the engine failed, but in reality the repair expenses were spread over several groups because that seemed like a fairer practice. The budget thus served as a way to represent an officially authorized practice rather than reality. Like the budget, this story of issuing work orders was a site in which the order of hierarchy was represented. That the two ways of proceeding exist simultaneously (one on paper and one in people's actions) was actually functional from the company's point of view because it allowed for work to be done effectively while established structures of power were still officially represented and thus were able to be invoked should the situation require.

For instance, Gary took me to meet one of the engineers for whom he usually conducted tests. Technicians worked for some engineers more often than others because the technicians usually worked in the same test cell and that test cell had a certain line of the company's products in it. Thus the engineers working on that product wrote orders for the same technicians repeatedly. This particular engineer said that he could send his work orders to the lab via computer. They would then go to Gary's supervisor, who would enter them in the normal scheduling process. However, this engineer preferred to print them out and hand carry them directly to the test cells so that he could talk to the technicians and explain exactly what he wanted. Thus his orders arrived on Gary's desk outside of the normal scheduling procedure, and Gary had to backtrack to get them on the schedule. He recognized the extent to which work orders needed to be supplemented in order to be effective. He did not entirely trust the text to serve as the sole representation of the technician's work, even though that's what work orders were officially supposed to do.

A similar departure from the official process occurred while I was observing another technician, Jim, modifying an engine by installing an experimental part. The engineer who had written this particular order came into the lab to see how the work was progressing and to ask Jim if he could rebuild another engine when he was finished with the current one. In effect, the engineer was trying to jump to the head of the priority list. A priority list of work orders was actually sitting on Jim's desk and needless to say, this engineer's request was not on it. When I asked Jim about the practice, he said that in order to avoid arguing with the engineer, he usually agreed to engineers' requests for special treatment but then did not always carry through on his agreement. In this case, he said he probably would do as the engineer asked because he needed a piece of equipment that he would get by working on the next engine (a sign that technicians sometimes needed to exert some effort to acquire the "capital" represented by tools and equipment that was shared among them). Indeed while I was there, he asked another technician who worked in a different area to prepare this second engine for Jim's work by removing a part. This technician joked, "One of these days we'll get an order so we know what's going on," and the technician who worked next to Jim joked back, "We don't need any orders."

This engineer and the technicians were functioning outside the official work order process with the goal of accomplishing a task that the engineer valued. This situation suggests both that the technicians did not need work orders to direct their work and that their work was more answerable to the desire of engineers for data than it was to the desire of managers for the maintenance of schedule and budget.

In addition to jumping the priority list, engineers worked around the system by getting their work order onto the priority list and then adding to it once it reached the technicians. That is, the engineers held their place in the line with a minimal order and then elaborated on the task as they waited for it to rise to the top of the priority list. This practice is evident in the last sentence of step 5 in figure 4.1: "Check with writer to determine whether additional FSC testing is required." The technicians resented this practice only because it could create problems with their supervisors. When the supervisor gave a work order to the technician, he or she would allow a certain number of hours for the job to be completed (coded into the work order format in the spaces for "start" and "comp[letion]" dates shown at the top of figures 4.1 and 4.2). If the engineer added to the order it would obviously lengthen the amount of time the order would take to complete and the supervisor could criticize the technician because the task was not completed in the allotted time. The organizational hierarchy within which the technicians worked called for them simultaneously to provide the artifacts and data that engineers needed, and to do so within the time frame their supervisor allotted. The need to take the time to provide sufficient data to satisfy the engineer did not always fit easily with the need to follow the orders of their supervisors.

In effect, the managerial text of the schedule sometimes conflicted with engineers' desire for data reports, the texts they valued the most. My observations suggested that in these situations, the engineers and technicians cooperated, as we see Jim doing, so that knowledge could be created even when the organizational hierarchy did not command it and indeed to some degree forbade it. Because the official work order process did not allow this kind of improvisation, it was done somewhat surreptitiously, a fact that allowed technicians' initiative to remain hidden and thus helped to construct and justify the corporate hierarchy. The only visible representation of the technician's activity was the work order, a text that represented the engineer's instructions as controlling the work to be done.

Thus the distribution of work orders sometimes happened in an improvised practice that existed under the orderly surface of the official work order process. The engineer-manager who coordinated between the laboratory and the engineers seemed to be undisturbed by such unofficial practice so long as test cells did not sit idle and there were no undue delays in accomplishing the work. He noted, for instance, that when an engineer took a work order directly to a technician this might be a sign that an engineering group that shared a test cell had decided its own test priorities for itself and made a reasonable change in a preset schedule. He thought of his job as involving *laboratory management* rather than *laboratory planning* because the former term suggested the need for flexibility. He knew that the formal written rules for distributing work orders sometimes had to be ignored if work was to be accomplished. Thus, when knowledge-creating work was being done in the Pacific Equipment labs, accomplishing work that led to knowledge seemed to be given priority over the organizational order of rules. Despite the frequency with which it occurred, however, informal working around was always treated as extraordinary, as a departure from order rather than a maintenance of it. Thus corporate hierarchy was preserved even as knowledge was made outside it. Work orders served as an orderly surface representation of a whole array of improvised activity.

SUPPLEMENTING THE WRITTEN WORK ORDER

The work orders, then, were meant to trigger work, but the relationship between work orders and the resulting work was not straightforward. To some degree, the engineers seemed to believe that following work orders should be unproblematic if the order was clearly written and the technician was competent. For instance, as I said in chapter 3, one of the engineers told me that he tried to write work orders in what he called a "cookbook" style because that way he could get the work done as he wished. In other words, the engineers sometimes seemed to think of the technicians as passive tools that the work orders should be able to activate. As we will see, there are problems with this way of viewing the laboratory technicians and their work.

In a discussion of the work involved in following instructions, R. Amerine and J. Bilmes (1990) say, "Successfully following instructions can be described as constructing a course of action such that, having done this course of action, the instructions will serve as a descriptive account of what has been done, as well as provide a basis for describing the consequences of such action" (326). In other words, people "following" the instructions do a great deal of creative and interpretive work to accomplish the actions that the instructions called for. They must, for instance, know how to read between the lines and do what is expected but not spelled out. They must know what is essential in the instructions and what may be modified if necessary to accomplish the same goal. They must be able to recognize when a step has been successfully completed. In the case of the laboratory technicians, they also had to be able to judge when the data they were generating were valid (as is evident in the second sentence of step 3 in figure 4.1: "Confirm correct engine operation with writer").

Glynda Hull (1999) has argued that blue-collar workers are handicapped in following instructions if those instructions are viewed as isolated steps divorced from the context that gives them meaning. She described a case in which a factory worker who was repairing circuit boards mistakenly removed a label that was necessary if the factory was to be able to trace the history of individual boards. The worker did not understand that customers required each board to be traceable and thus did not understand the purpose of the label. Hull attributed this lack of understanding to the way in which the company she was observing conceived of an appropriate division of labor, one that failed to provide the factory worker with the information he needed to do his job. To some degree, Pacific Equipment conceived of the division of labor similarly, seeing the technicians' work as primarily manual work that followed a "cookbook" set of instruction. However, the Pacific Equipment technicians whom I observed actively sought to understand the situation surrounding their work, and that understanding was probably necessary for their successful completion of work orders. For instance, Rich told me that engineers sometimes wrote orders requiring technicians to build or run a vehicle in a nonstandard way that might seem foolish to a mechanic, but that technicians understood that the engineer was often experimenting and that the goal was sometimes to learn something rather than to create a permanently usable part. Thus Rich understood not only when to apply the expertise he had gained by experience, but also when not to. This kind of delicate balance is not achieved by functioning like a tool that blindly follows orders. It is highly creative work. However, once the actions that work orders call for are accomplished, that creative work fades from view and only the instructions remain. The course of action then looks like the instructions controlled it from the start and the actor looks passive. Thus, as a genre, the work order both triggered and concealed technicians' work.

Indeed the fiction of the stand-alone work order concealed necessary work by the engineers as well. The written work orders had to be supplemented and supported by efforts from everyone that were oriented around the work order and disappeared into it. For instance, although engineers could talk about the desirability of making the orders complete, in practice completeness was impossible because the writer could not anticipate all contingencies. The engineers' desire to talk to the technicians shows one supplement that all said was vital. Everyone I talked to about the process of following work orders said that the technicians and the engineers needed to be able to talk to one another, sometimes face-to-face and more often by telephone. The expectation of such communication is evident in steps 3 and 5 of figure 4.1. Face-to-face communication normally took place when the engineer visited the lab, such as we see an engineer doing at the end of vignette 4. Such communication was important because it allowed engineers and technicians to build a working relationship, something both Rich and Gary told me was important.

Moreover, being together in the lab allowed the engineer and technician to look together at the engines, test stands, and parts they were working with. Such examination is particularly useful in an engineering development center because nonstandard parts and configurations are often being experimented with so the writer cannot always rely on the technicians' familiarity with standard practice to supplement the written words. For instance, I observed Jim rebuilding an engine when the engineer who had written the order visited him. The part lying on Jim's workbench waiting to be installed turned out to be the wrong one. The engineer spent about fifteen minutes poking through parts in the room looking for the right one before concluding that it was not there and going to get it from elsewhere. Jim told me that the part to be used was not specified on the work order, so he had planned to use a standard part rather than the experimental one the engineer wanted. Jim was working on this rebuild only because the technician who normally worked for this engineer was on vacation. He speculated that the part was not specified on the work order because the engineer had orally explained what he wanted to the technician who usually worked for him or because it had been specified on the last ten work orders and thus would have been part of the usual technician's expectations.

Engineers and technicians also maintained contact by phone, as we see Rich doing in vignette 4. Gary told me that at one time the only phone in his area was in the supervisor's office. Now there was one in every test cell. The easily available phones made for less control by the supervisor but for more effective work. The company had opted for the latter, another instance of tension between the order of effective knowledge-generating work and the order of hierarchy. Pacific Equipment had apparently decided that rigid control of the technicians' actions reduced the effectiveness of the technicians' work, despite the fact that the company's understanding of work orders still represented the engineers as controlling the technicians.

In addition to being supplemented by contact with the engineer who wrote them, the work orders were supplemented by other kinds of writing as well. These might include service manuals, standard instruction sheets, marks on the parts, drawings, and even photographs. For instance, I watched Jim working on an order that called for him to change the pistons and liners in an engine so that the engine's performance with the new parts could be compared to its performance with the old. He told me that the engine was a French-made Pacific Equipment engine that he did not work on very often, so he was using the company's service manual in conjunction with the work order. Under these circumstances, he was unable to do what he would ordinarily do and follow lines on the work order that read, for instance, like step 3 in figure 4.2: "Remove all of the pistons and replace with the part number listed below." He had to look up or at least check the steps in the manual to be sure he was doing them correctly. When he began to install the experimental parts, the manual of course contained no reference to them, so he also used a set of written instructions that came with the parts. I also saw both him and Rich look away from work orders to consult photographs they had taken of parts they had built previously so that they could duplicate work they had done before. Rich also told me that when work orders provided what he considered to be very minimal diagrams for building an electronic part, he made up his own charts and tables. He did not mind doing this because he knew the engineers "had engineering to do" and he could do the charts for himself.

As Rich's constructing charts suggests, in addition to supplementing their work orders by visiting the technicians or phoning them, the engineers were able to rely on the technicians' expertise in familiar situations to shorten their own writing task. They did not have to spell out every action because the technicians would already know what to do in standard situations. So in vignette 4, Rich uses one work order that simply asks him to repeat work that he has done before without specifying any details. Indeed, the engineers had to rely on the technicians whether they wanted to or not because they could not anticipate contingencies such as missing parts (or, as happened in one instance I saw, a nest of wasps in an engine) and so had to rely on the technicians to order these unexpected events. Thus almost all work orders contain a line such as those in step 6 of figure 4.1 and step 4 of figure 4.2: "Do other work as needed" or "Perform any additional work to complete" or (my favorite) "Please try to anticipate any future problems with this project." One technician who was not in the study told me that he had never read a work order that actually told him what to do. He said that they all read "Do as needed." In one way these sentences reveal the tension between engineers and the technicians in that the engineers saw these sentences as defensive; they believed that if they did not have such a line in an order, technicians might refuse to do anything that was not specified. But these sentences also reveal a great deal of reliance on the capabilities of the technician and do not at all reflect a cookbook notion of their work. As Jim said, "They assume we know a lot." The technicians told me they resented this only because they did not get credit for the extra work time such a sentence might involve. As Shapin (1989) noted was the case for Boyle's laboratory assistants, they also did not always get credit in the official recognition of how work is done.

Even in small ways, the laboratory technicians I observed had to supplement the information in work orders. For instance, when Gary showed me what he said was a typical work order, he pointed out areas that he saw as potential problems. For example, the order called for a "wet/dry" test to be done. Gary told me that this test was no longer done but had been replaced by readings done on a dew point meter. The engine number on the work order did not match the one taped to the engine in Gary's test cell or the one engraved on it, which also did not match one another, but Gary said he knew this was the engine to be used because he had had it in his cell for a long time. The brand name pump mentioned in the subject line of the order was different from the brand name mentioned in the order's first line, a variation Gary attributed to the engineer using a previous work order as a template for this one and failing to change the first line. Gary assumed that the pump mentioned in the subject line was the correct one, but he phoned the engineer to make certain. Thus even this relatively good work order required effort from Gary to make it effective. He said that some technicians followed the work order "to a T," but that he had been around long enough to know that this was not always the right thing to do: "If you do everything on that order just the way it's written down, chances are something won't work out. They can't foresee all the problems you might have."

So the written script of a work order is a kind of fiction describing a simple, logical sequence of actions that the engineer chooses and that the technician follows. In reality, however, orders are not simple things. Rather, to be effective, they must be supported by a whole system that includes oral interaction, texts, and the technicians' expertise. It is only after the work has been accomplished that these supports are forgotten and the work order seems to stand alone as the description and cause of the action. At that point, much of the knowledge-generating work that is unique to the technicians vanishes and only the engineer's planning seems to remain. Thus it was common for engineers to talk about laboratory work in ways that appropriated it to themselves in quite natural ways. "I didn't change those return springs when I lowered that pressure," said one at a meeting, meaning that he had not asked the technician to change the springs when the technician made the adjustments necessary to lower pressure. "I need to remove some teeth from a gear," said another as he wrote a work order for a technician to do just that. And a third engineer asked a coworker, "On the traces you ran, . . . did you have anything plugged into the pilot valve?" The traces were, of course, run by a technician who would have set up the pilot valve. The engineers' language treated the technicians' actions as belonging to the engineer even though he or she was not the one who physically performed them. This ownership was established by means of the description of work appearing in the work order.

Thus, despite their status as semi-tools, the technicians actively participated in the creation of the hierarchy in which they worked. They built and tested the devices designed by the engineers. They cooperated in working around the rules for how work orders were to be distributed without openly challenging those rules. They contributed their expertise (their cultural capital) to carrying out work orders, and, in the part of this process they were least happy with, they allowed their contributions to be seen simply as following the engineers' instructions, so that credit for their work often rested with the engineer who wrote the work order. In their large and small daily activities, they helped to create and maintain the hierarchy of the Pacific Equipment engineering center and to see that work was accomplished. Thus they participated, as Foucault says we all do, in the workings of power: "In reality, power in its exercise goes much further, passes through much finer channels, and is much more ambiguous, since each individual has at his disposal a certain power, and for that very reason can also act as the vehicle for transmitting a wider power" (1980, 72). Work orders were plainly one means by which power worked through the relation of the technicians and the engineers to the advantage of the latter.

TECHNICIANS' TEXTS AND DEFINITIONS OF GENRE

Work orders, then, were texts that ordered and defined joint activity across discontinuities within the engineering center. In contrast, within their own workspaces, the technicians also produced texts that affected the shape of this organization but the effect of these texts was confined to the laboratory. Because the technicians' attempts at textual ordering were mainly confined to their own work and workspaces, they were largely carried out without interference but also without support from above. Their texts were less visible than work orders within the overall organizational activity and less the focus of organizational attention. As a consequence, the organization was less aware of the technicians' literate activity than it was of the engineers'. Moreover, the limited visibility of the technicians' texts meant that the organization did not perceive them as generic, a category that is more political than commonly acknowledged in theoretical discussions about genre, which tend to treat it as a rather neutral concept (Miller 1984; Berkenkotter and Huckin 1995).

In the lab, technicians made notes to themselves as they worked in order to shape their actions. These notes could be as simple as check marks next to steps they had completed or as complex as the wiring diagrams that Rich generated for his own use. Rich also put up numerous signs to help himself maintain order among the hundreds of parts he maintained in inventory. Thus he labeled everything from shelves and baskets of parts to individual wires in the parts he created. He also used signs to try to shape the actions of others who were in and out of his area all day delivering or picking up parts. Thus he marked one area as the Setdown Area where a particular vendor was to leave deliveries. A sign taped to his desk read There's Nothing Here that Belongs to You, So Leave It All Here!!!, a warning he said was necessary because colleagues did not always distinguish between parts that were in inventory and could be borrowed and the possessions on Rich's desk. Another sign over a rack containing spools of wire read If You Are Not Intelligent Enough to Remove and Replace the Rubber Bands that Hold the Wire from Unwinding—Or You Have a Handicap that Prevents You from Replacing Wire Spools When They Are Empty—Then You Shouldn't Be Using Wire from This Rack!!!! Thank You. Rich told me that this last sign was new, as was evidenced by the fact that no one had yet attached any notes responding to his obvious annoyance. He told me that people were actually quite responsive to these signs. For instance, he had taped a sign on a countertop asking people to put returned parts away rather than to pile them on the counter and had not had an accumulation of parts on the counter since then.

Do technicians blame for mistakes or engineers take?

One of the characteristics of these technician-generated texts is that within Pacific Equipment they were perceived as more improvised and idiosyncratic than work orders. It would certainly be possible to demonstrate that there were also idiosyncratic features in work orders generated by different engineers, but no one in the company would say that in writing a work order an individual engineer spontaneously created a textual form in response to a unique problem. That is, work orders were perceived as belonging to an institutionalized genre, as texts whose form was shaped by the organization as a whole and behind which the authority of the organization lay.

In contrast, the texts that the technicians created were seen as shaped by individuals. Rich's signs, for instance, were very much seen as his, which was why his fellow technicians enjoyed attaching provocative notes to his written warnings (a kind of activity that is inconceivable in response to work orders). Thus it is possible to claim that the texts produced by technicians were less generic than the work orders. At Pacific Equipment, the technicians' texts were confined to the lab and did not circulate widely. They stayed within the field of the technicians so that there was no need to shape them to suit the needs of more dominant groups. Moreover, the work they accomplished through those texts was largely made invisible according to how the dominant genre of work orders was understood to function. Thus, one could understand Rich's cautionary signs as part of an organizational genre. Indeed, they did echo a company practice of using signs to order space and action. The floors in the hallways, for instance, were painted with lines that indicated where one could safely walk, and doors were marked with warnings that hearing protection should be worn beyond a certain point or that only authorized people could enter. But Rich's texts were not considered part of the same family. Rather, they were seen as originating with him individually and therefore lacking the power that comes from institutional backing. The nongeneric status of technicians' texts suggests the corporate invisibility of their literacy activity, whether it be writing signs or reading and interpreting work orders. Thus it reflects and perpetuates the way in which their work disappears into the corporate effort and thus reflects and perpetuates the hierarchical order of the Pacific Equipment engineering center.

POWER AND GENRE

What then does this chapter contribute to our understanding of the relationship between text, genre, and the creation of a power structure? If Pacific Equipment work orders are not idiosyncratic, then some texts in organizations are tools around which the activity of different groups and the way in which the activity is understood can be oriented. At the company, I have shown that these texts would include budgets and documentation as well as work orders. It is quite possible that such texts have generic status within the organization. Indeed, they are quite likely to have been institutionalized as required textual forms. That is, power relations and perceptions of work lead to, and from, perceptions of genre.

Such texts can draw together the work of discontinuous areas in an organization. They can also shape the form in which this is accounted for and understood. In the case of work orders, texts shape technicians' actions. While the technicians ultimately control their own activity and the contribution it makes in the engineering center, work orders, read retroactively, give the appearance of control to engineers. Thus they shape participants' sense of the relationship of the work of the engineer to the work of the technician, privileging the work of the former. Moreover, because texts such as the work orders are institutionalized as genres, they encourage the perception and enactment of a consistent order and discourage looking at activity in different ways. They help the social system's participants to see it as ordered in a certain way, and participants then act in accordance with the order they perceive. That is, these texts do political work; they create and reinforce positions of power within the organization.

The connection between the status of genres and the status of writers is implied in studies such as Schryer's (1993) in which inhabitants of a veterinary college privileged either the scientific research article or the veterinary medical records system. The status of the genres and the status of the genres' users were plainly intertwined, with clinicians and their genre receiving less respect than researchers and their genre. Similarly, Beverly Sauer's (1998) description of how officials ignored some testimony about potential dangers in a coal mine allows us to see how the ignored testimony did not fit into an identifiable genre, a fact that in turn reflected the low political status of those giving the testimony.

Genre theory tells us that workplace literacy is more than a matter of knowing how to read and generate grammatical prose. A successful writer must also be able to "read" the typified social situations that indicate which kinds of text are appropriate. This individualized notion of literacy is not enough, however. In the workplace, writers function within organizations that value some work more than other and that increase or lessen the likelihood that they will produce texts that others will heed. As Bazerman points out, "The subordination and division of laboratory labors as well as the participation of individuals in the aggregation and distribution of collective work are realized through the discursive spaces each member of the collective can come to inhabit, in negotiative dialectic with other members of the collective" (1997, 304). The organization of a social system constitutes some tools and not others as useful and enables some actors and not others to use them (Hull 1997, 1999). Textual genres are among the tools that are so constituted and made available. Literacy is a complex system in which an individual writer fits into an ecology from which she can draw strength (or not) and on which she can have an effect (or none). The social systems in which blue-collar workers function may be one of the factors that leads to their being considered less literate than white-collar workers, because opportunities for, and definitions of, literacy reflect the work of the dominant group. Thus much work gets hidden or lost or taken for granted by the dominant group. Granting or denying generic status to a group's texts is one means that an organization can use to signal whose words are authorized (Bourdieu 1991, 111) and thus who will be granted the symbolic capital that such authorization provides.

For blue-collar workers, such authorization is largely unavailable. However, for another group of employees, summer engineering interns, the authorization that comes with an engineering position is provisionally granted even when the intern is struggling to do the work for which he or she has been hired. Vignette 5 shows an intern puzzling his way through various tasks with the help of the cognition that Pacific Equipment has distributed among his fellow employees and the tools that are made available to him.

Vignette 5

An Engineering Intern's Morning: Kevin

I arrive at the engineering center a few minutes before 7:00 A.M. Kevin, an intern, picks me up from the security desk as he comes in to work and I follow him through the maze of cubicles until we reach his. He immediately boots up his computer and then opens his planner, in which he has made his to-do list. His computer first displays the company's news of the day. He skims this briefly and then checks his E-mail. He tells me that he is designing two things for an off-highway vehicle: brackets to hold a toolbox and parts such as locks for the gas cap that are supposed to help prevent vandalism.

He takes an engineering drawing from his desk and then goes to another cubicle where he shows the drawing to an engineer named Brian. Brian says that he had talked to someone in marketing about how many options the factory should offer. He knows that tests are currently being run on the toolbox brackets Kevin has designed and asks how much the toolbox weighs. He thinks the vehicle with the toolbox is currently in the sound room because the technicians were trying to figure out a source of gear whine. Kevin wants to check on the brackets, so he asks for directions to the sound room. Brian is not actually sure of the best route to take because renovations are currently underway in the lab (indeed, these are the renovations that Mark is working on in vignette 1), and those renovations block the route he would normally take. He suggests that Kevin just ask someone to take the toolbox off the vehicle and bring him the parts. The engineer in the next cubicle suggests that Kevin check for any rattling the toolbox might be generating because customers might object to such a noise.

Back in Kevin's cubicle, he opens a spreadsheet listing engineering drawings he's working on. He also gets out a paper file of drawings. He begins entering dimensions on an on-line drawing. He consults the paper file of drawings so that he can replicate a note he made on a previous drawing. After he has worked on his drawings for a while, he wants a break and decides to go to the lab and to check on his toolbox brackets to see if they have fatigued or flexed. We pause outside the lab doors so that I can don protective gear and then we plunge into the large laboratory area.

Kevin says that he seldom goes to the labs and, since this is his first year as an intern in the engineering center, he has trouble finding his way around. We hunt for the sound room and eventually find it, but the vehicle he's looking for is not there. We look outside the building in an area where many Pacific Equipment and competitor vehicles are parked but still don't see the one he wants. We go back into the building and he asks Brian where he thinks the vehicle might be. Brian says he's not sure but suggests that another engineer might know. He consults this engineer, who says the vehicle is in a test cell.

Kevin returns to his cubicle to work on his monthly report for his team leader. Because he is a student, the team leader had told him that his report could be brief. Kevin calls up the previous month's report and adds a few sentences directly in the text of the old report so that his text is automatically formatted to match the older report. Then he erases the rest of the old report and sends this to his team leader.

He consults his spreadsheet list of drawings again and then opens another drawing and begins to work on it. His phone rings. The call is from the technician who is building a part that Kevin had designed. The technician doesn't have enough material to build the part out of one piece of metal and wants to know if he can weld two pieces together. Kevin calls up the drawing of the part the technician is talking about. "That's our critical corner," he says. "That's where we get all our fluxion. Is there any way you can load it above that?" He and the technician reach some sort of agreement and he hangs up.

Kevin checks his E-mail again and then we go to the cafeteria for a break and then to the laboratory to search for the test cell containing the vehicle with his toolbox. We find a test cell containing the same kind of vehicle but this vehicle is not the one with Kevin's toolbox attached.

Back in Kevin's cubicle, he again works on his drawings. He checks a company parts catalog, telling me that another vehicle group already has antivandalism equipment on its vehicle and when he can, he uses their design and adapts it. He needs a part number for the cap screw he wants. He can't find what he wants in the catalog, so he consults an on-line database with the same information. He still can't find what he wants, so he asks an engineer across the aisle from him if he knows how to look up the part number. This engineer consults the catalog and then tells Kevin how to use the on-line database, walking him through the process step-by-step. Unfortunately, the process doesn't work, so the engineer asks Kevin to "let me play around with it for a while" and returns to his own cubicle to try to figure out the process. Eventually, he comes back and explains how to use the database. Then he observes that the screw Kevin was looking up is "a pretty good size. Do you need that?" Kevin explains why he thinks he does.

Kevin once again asks a coworker where he thinks the vehicle with the toolbox might be. This coworker gives Kevin a phone number that he can use to check on the toolbox vehicle's whereabouts. Kevin calls but whoever answers the phone has no notion of where the box is. Kevin goes back to Brian and says that he found the right kind of vehicle but the toolbox wasn't on it. Between the two of them, they figure out that there are two identical vehicles and Kevin needs the other one. Kevin goes back to a second coworker and asks about this second vehicle. The coworker says that it is on the test track outside the building. Kevin decides to go to the track. Another engineer says that he will be going too because he wants to drive the vehicle. He's in the transmission design group and he wants to check out noise and handling.

We go outside and drive a competitor's vehicle to the test track. Kevin tells me that Pacific Equipment encourages employees to drive competitors' vehicles because "it gets you motivated." We finally find the right vehicle. It's being driven around the test track for a hundred fifty hours. Kevin hails the driver who stops near us. Kevin measures the clearance below his toolbox to see if the box has sagged and also pulls on it to see if it's loose. After we finish looking at the toolbox, Kevin drives me around a larger test track to show me how fast this competitor's vehicle is. "This is the part I like about work," he declares.

We return to the cubicles. "I found it," he declares to Brian. He tells him that the clearance is still good and that it took a lot of force to move the toolbox. Brian says that this test is a good one because it lasts for so many hours. "Too bad it's on that nice smooth track, though," he adds. Kevin returns to his own cubicle and makes a note in his planner. Then he begins to follow the instructions he has been given for using the on-line parts catalog.

Around eleven o'clock, it's time for Kevin's lunch break. He escorts me out of the building on his way to the cafeteria.

CHAPTER 5

ENTERING SYSTEMS OF KNOWLEDGE/POWER

I think school is good and all, but I think [Pacific Equipment] cares more about the fact that you made it through those four years of mechanical engineering or whatever because that way they know that you can learn.
—Pacific Equipment summer intern

. . . learning is so fundamental to the social order we live by that theorizing about one is tantamount to theorizing about the other.
—Etienne Wenger, *Communities of Practice*

One way to make knowledge and power practices more obvious to outsiders is to follow newcomers as they are learning about them. In the last few years, we have seen a number of interesting studies of newcomers learning to write in the workplace (e.g., Freedman, Adam, and Smart 1994; Freedman and Adam 1996; Winsor 1996; Katz 1998, 1999; Beaufort 1999; Dias et al. 1999; Dias and Pare 2000). This work draws on theories of situated activity to show that actions always take their meaning from the situation in which they are performed and that they, in turn, help to constitute. However, this otherwise useful research has not yet examined in any detail the way in which hierarchical workplaces require newcomers to learn to operate within structures that are simultaneously those of hierarchy and of distributed cognition. We do see implicit attention to these issues in works such as Jamie MacKinnon's (1993) study of how newcomers were taught to write by their superiors at the Bank of Canada or Natasha Artemeva and Aviva Freedman's (2001) study of how,

Education

125

among other aspects of culture, corporate hierarchy affected work practices at a computer company. However, none of these studies focuses explicitly on the interaction of hierarchy, knowledge, and texts. At Pacific Equipment, newcomers must learn to operate within interacting structures of power and jointly constructed knowledge, using conventional texts as a means of entry.

In this chapter, I look at six engineering students who worked at Pacific Equipment during the summer of 1999 and explore how they simultaneously gained access to knowledge and power. The company hired about 20 of these interns each summer, beginning when they were sophomores and continuing throughout their college years; if all went well, they would become permanent employees when they graduated. In examining their experience, I draw upon theories of distributed cognition to examine how newcomers enter into distributed cognition in workplaces where what counts as valid knowledge is constantly changing and where the validation of knowledge is at least partly effected by textual maneuvers meant to position oneself within systems of symbolic power. If we think in terms of distributed cognition and changing knowledge, then learning to write in the workplace entails getting plugged into a system of endless knowing rather than gaining a bounded set of ideas. If we think in terms of a link between knowledge and power, then learning to generate knowledge also means appropriating the available means to power.

Given the negotiations around knowledge and around power that I have described in chapters 2–4, consider what engineering students must learn in order to function successfully at Pacific Equipment. They must be able to fit whatever general engineering knowledge they have learned in school with the situated expertise about vehicles that Pacific Equipment engineers have amassed over the years. They must also work jointly with others to change that knowledge so that the vehicle is improved. In vignette 4, we see John and his colleagues operating from this mix of given knowledge and negotiated, provisional understanding generated by new data to which they have not yet assigned a meaning. Newcomers must learn to negotiate this knowledge with fellow engineers and to defend it to managers, as John and his colleagues evidently anticipate doing when they talk about funding for their project. Newcomers must also be able to draw on the expertise of the technicians but establish their own right to determine the meaning of whatever data the technicians generate. None of these tasks is easily learned in school.

The experience of the Pacific Equipment interns suggests the way in which both power and knowledge result from positioning. I have already argued that power is established partly by means of the relative positioning of those involved, so that managers acquire the right to issue budget, for instance, or so that engineers can issue work orders not because of some personal power, but because of the role to which the company has assigned them. To quote Michel Foucault:

> The idea that there is either located at—or emanating from—a given point something which is a "power" seems to me to be based on a misguided analysis, one which at all events fails to account for a considerable number of phenomena. In reality power means relations, a more-or-less organised, hierarchical, co-ordinated cluster of relations. (1980, 198)

At the Pacific Equipment engineering center, these relative positions already existed when interns arrived to take up their summer jobs although, of course, people filling a position necessarily modify it (see, e.g., Katz's 1999 study of newly hired writers shaping their own positions). The interns were intended to fit tentatively into the slot marked Engineer even though they had not yet completed their engineering educations. An engineering position gave them the social capital or status that entitled them to do certain things, such as issue work orders, assuming they could figure out how to do so in an acceptable way. An engineering position also gave them access to more experienced engineers and to engineering tools, including discursive ones. In other words, it gave them access to the distributed cognition already operating within the engineering center. Being identified as an "engineer" at the company thus positioned them within an institution of the sort that Barry Smart says uses "methods for the formation and accumulation of knowledge . . . as instruments of domination and [uses] increases in power . . . to produce additions to knowledge" (1985, 91).

In writing about newcomers' entry into situated activity, Etienne Wenger points out that assuming a place in such activity affects the identity of the newcomer: "Because learning transforms who we are and what we can do, it is an experience of identity. It is not just an accumulation of skills and information, but a process or becoming—to become a certain person or, conversely, to avoid becoming a certain person" (1998, 215). Wenger points out that such a change in identity requires an interplay between "identification," which he defines as "an investment of the self" (188), and "negotiability," which he says is the "degree to which we have control over the meanings in which we are invested" (188). In other words, newcomers have to accept the role that the organization has defined for them, but they also have to reshape that role so that it fits with their sense of themselves. Wenger claims that talking about identity in this way necessarily involves discussions of power (189). If a newcomer simply accepts the meanings that already exist in a community, says Wenger, the experience is one of "powerlessness—vulnerability, narrowness, marginality" (208), not only because the newcomer is subject to the meaning imposed, but also because she is not able to contribute to the generation of new meaning or knowledge that the community might value. On the other hand, if the newcomer simply rejects the existing meaning and knowledge, then she experiences "meaningless power, freedom as isolation and cynicism" (208). As I will show in this chapter, the Pacific Equipment summer interns took up already established practices and meanings to shape their emerging identities as engineers. Simultaneously, they learned to contribute knowledge of their own by using the means, including the textual means, that the community of practice had made available.

I will be discussing six summer students in this chapter: Mark, Kevin, Don, Linda, Sue, and Jane. The men were all seniors. Linda was halfway through her sophomore year, and Sue and Jane had just finished their first year of college. In this chapter, I first examine how genres helped distributed cognition at Pacific Equipment to function and helped the interns to establish positions within it. In other words, I begin by talking about how the interns gained access to some of the knowledge already in play at the engineering center and then how they were able to participate in the generation of new knowledge. Among other things, I argue that the interns' experience shows that creating knowledge is a less purely cognitive activity than we sometimes assume. Edwin Hutchins (1993, 1996) demonstrates that cognition is distributed, not only among coworkers, but also in their tools. I will argue that one of those tools is genre itself and that, moreover, genres are stabilized partially by the software tools used to generate them. Genres were one of the communication tools that Pacific Equipment employees used to access currently stabilized knowledge and to generate and stabilize new knowledge (cf. Freedman and Smart 1997). Because cognition had been distributed into tools, including genres, as well as people in the engineering center, learning to use genres (and the software used to create them) was one of the factors allowing even these novice employees to participate in generating knowledge.

After discussing how learning to use genres helped the students to learn and generate knowledge, I will discuss how learning to use various genres to which engineers were entitled also positioned the summer students as engineers, and thus allowed them to exercise the power that comes with this position at Pacific Equipment, although depending on the amount of education and experience the students had, their tasks were often the less crucial ones (cf. Lave and Wenger 1991 on legitimate peripheral participation). The students' experience shows how genres interacted with both parts of the knowledge/power dynamic in the engineering center.

USING GENRES TO ACCESS DISTRIBUTED COGNITION

As Wenger suggests, one aspect of assuming a professional identity is becoming acquainted with the knowledge that already exists, knowledge that he says has been "reified." Wenger says that he uses "the concept of reification very generally to refer to the process of giving form to our experience by producing objects that congeal this experience into 'thingness.' In so doing we create points of focus around which the negotiation of meaning becomes organized" (1998, 58). In the engineering center, texts were important objects for the reification of knowledge. In texts ranging from data curves to technical reports, engineers solidified their current understanding of the vehicles on which they worked. For the newcomer, the reifications of knowledge that are already in place are "essential to repairing the potential misalignments inherent in participation" (65). That is, until newcomers learn existing, reified knowledge, they will be powerless to contribute new knowledge because they will be unable to bring their actions into alignment with those of relevant others. Presumably, school is one of the places where new-comers learn a discipline's reified knowledge, but the very existence of internships suggests that school is inadequate by itself. That is, in places like the engineering center, the knowledge that counts includes the local understanding of the company's specific products. Without it, any engineer working there is almost useless, no matter how solid his or her engineering education has been.

At Pacific Equipment, the interns ran into particular problems in accessing existing knowledge because the knowledge the engineers worked with was often fluctuating and uncertain. One intern's supervisor commented on the contrast between knowledge at the engineering center and knowledge in school: "At school, if you don't have the answer to a question, you get no credit. That's not the case here. Half the stuff we do here, we don't understand." The ability to tolerate such uncertainty was crucial because, at the engineering center, the whole organization was learning, as well as any individual. Current thinking in management circles argues that organizations engage in learning all the time (Nevis, DiBella, and Gould 1995; Cohen and Sproull 1996). However, in an increasing number of worksites, new knowledge is the primary product. As I said in chapter 4, an engineer once told me, "We make two things at the engineering center; we make the drawing and we make the spec[ification]." The company's factories produce its products, but the engineering center produces only knowledge about those products, and even that knowledge is often about planned products that do not yet exist and indeed may never be built (cf. Medway 1996 on documenting virtual buildings in architecture). Thus knowledge was always uncertain.

While the ability to tolerate uncertainty is important, engineers can do so for only so long before they have to decide how to proceed with their design work. Thus they seek ways to reduce uncertainty when they can. For instance, we see engineers wrestling with uncertain knowledge in vignette 4, where John and his colleagues don't understand what their data mean. John had tried to solidify (or reify) their knowledge by creating a graph in which he extrapolated from existing data to project what will happen with changes to their product that have not yet been enacted. However, the uncertainty here was too great and his attempt was not success-ful, at least for the time being. In the face of unstable knowledge, Pacific Equipment engineers needed tools that would allow them both to access knowledge that had already been agreed upon and to stabilize knowledge that was currently in flux long enough for it to be shared, evaluated, and analyzed. One means for doing both of these things appeared to be not just texts, but generic texts such as the graph John displayed, which attempted to extend an existing form to shape his and his coworkers' understanding of their data. In this study, conventional texts were a means for accomplishing sta-bilization because their repeated form guided a writer and encour-aged repetition of previous ways of thinking even when the writer did not fully understand the previous thought. That is, knowledge existed not just in the concepts in people's heads but also in the tools they used,[1] including genres. Using genres helped writers to code their understanding in ways that coworkers would more easily recognize and accept; it also helped to shape writers' understanding of their work as the repeated form exerted pressure upon how con-cepts could be expressed and thus reified. Indeed, ideas that had been reified were almost always expressed in a generic form. The importance of genres meant that the Pacific Equipment interns devoted effort to learning them.

A genre, as Carolyn R. Miller (1984) tells us, is a typified response to a typified social situation, and the typified situation is usually thought of as the more essential of these elements. However, a genre's form, the conventional language and organization it uses, is often the means by which its presence is being signaled and by which the meaning of the genre is being invoked. To the Pacific Equipment interns, genres often appeared first simply as conven-tional language because they did not always understand the com-plex, historically shaped social situation to which the language responded (cf. Winsor 1990b). However, learning conventional language was a step on the way toward understanding the social sit-uation (Vygotsky 1978) and, just as important, toward fitting them-selves into it as engineers and responding appropriately. Pierre Bourdieu (1991) argues that language has to be used by an appro-priate person in an appropriate way in order to have a social impact. For the engineering interns, learning conventional language meant that they were learning reified knowledge while simultaneously learning to sound like, and thus be heard as, engineers. Thus in genres, knowledge and power were very tightly joined.

The most extended instance of learning conventional language that I observed involved Linda in a conference with her supervisor, Ben, who was responding to a set of drawings that Linda had prepared. Drawings are an important engineering genre, and Ben spent two hours helping Linda to execute hers appropriately (cf. Katz 1999 on the importance of writing reviews in helping newcomers learn the social complexities of their workplaces, and Blakeslee 1997 on the impact of discussions between experienced and novice authors of a scientific paper). When Ben helped Linda put the proper written explanations—what engineers call "notation"—on her drawing, a large part of what he did was to push her toward being more conventional. Thus he made comments like "It's preferred not to dimension to hidden lines" (meaning that the length of a part's side—its dimension—should be indicated relative to a line that showed on the drawing rather than to one that was hidden because of the angle of the drawing). Ben relied on these language conventions even when he did not understand the thinking behind them. He examined a remark on the old print, for instance, and said that he assumed it was in a standard form. He said that he didn't know why the standard was the way it was but that "I would just copy it." This reliance on the model was repeated throughout his session with Linda: "Let's do it like they did and call out datum B as one side" (meaning that they would use one side of the part as a reference—a datum—from which the distance to the other would be measured); "We'll use the same note they did about true position" (the position of a part relative to a fixed point on the drawing); "Let's dimension this the way they did."

One of the things happening in the conference between Ben and Linda was that the marks on Linda's drawing were moving further from standard English and closer to specialized engineering language and writing. For instance, on her drawing, she had incorporated an English sentence saying that the holes around a circular part should be "evenly spaced." Ben acknowledged that her phrasing would tell the builders what she wanted, but he told her that such a phrase was not conventional. He then showed her how to specify the same information using angles and geometric terms. At another point, he indicated a place on the old print that read "index not important," and said "That's the way to say. . . ." something that she had taken three lines of English to say. At still another point, he changed the word *deep* that she had used to a symbol of a short line with a downward arrow attached to it.

Using conventional language could serve multiple functions: economy, for instance, as a symbol came to stand for a whole sentence or concept that was assumed to be part of people's disciplinary knowledge and thus not in need of explanation. Moreover, using this conventional language would help Linda to establish her identity as an engineer, rather than as an outsider, in the eyes of other people at Pacific Equipment. The correct use of disciplinary genres was one of the abilities that would mark her as a member. Most importantly perhaps, there is also a strong element of distributed cognition in the use of conventional language and models. For instance, Ben spoke of duplicating the tolerances on past drawings as a way to "leverage" the experience of past designers and production engineers. Unless he had reason to do otherwise, he assumed that the tolerances on the drawings reflected the experience and knowledge built up during previous design work. He did not need to understand the rationale behind the tolerances because he trusted that the ones in the previous drawing had proved acceptable in the past. This knowledge was coded into the drawings even if he was not able to see past the code (cf. Winsor 1990b). Written documents—in this case, a drawing and its notations—reified knowledge, at least temporarily. Using engineering language meant that Linda had use of the knowledge that Pacific Equipment had built up and coded into its parts and drawings and into other records and symbol systems. It also helped her to establish her identity as an engineer with her coworkers. The conventions making up the genre of the drawing provided reified knowledge for both Ben and Linda.

Like Linda (and Ben), the other students, too, were sometimes able to make use of the knowledge coded into conventional language even without fully understanding it. Of particular interest to me were some Vygotskian moments in which using conventional language forced students to use words before they fully understood them (Vygotsky 1978, 32). Language was invested with meaning in action. For instance, Sue told me that "I have PLC equipment here, and I had no clue what that meant. The computer guys were like 'well, is it PC or PLC?' And I'm like 'I don't know' and finally I realized that PC is just the computer and PLC is just the test stuff or whatever, so now I know what it means." Having to distinguish between the terms forced her to distinguish between types of equipment.

Similarly, before Linda had the meeting with her supervisor to discuss her design drawing, she told me that the meeting should be "interesting" because she didn't always know what she wanted to say and thus she and her supervisor would have trouble communicating. In our interview, I asked her to clarify this statement; was it a problem she had communicating in general or specifically in this situation? She explained that she meant the claim to apply in a "more technical sense. I hadn't had a lot of experience in what I was doing at that time with the couplings. There's a specific kind of tolerancing and dimensioning we use called GD and T, and I haven't used that before. . . . It wasn't clear to me what I needed to be doing to be able to ask the correct questions." Her language was limited by her limited experience. And she clearly valued learning language: "This is my third session here and this has been the best as far as someone teaching me things that I'm unfamiliar with, teaching me terms."

In yet another example, Don also ran into terminology problems regarding a design drawing. One of his coworkers told him to call someone building the part he had designed and to give them directions using the terms *one-inch fine* and *NPTF*. The following brief exchange makes it clear that Don did not understand these terms even though he had to use them:

Don: One-inch fine? They will know—?
Coworker: NPTF.
Don: NPTF? That's like a standard and they'll know what
    I mean?

In these instances, the language preceded the knowledge rather than the other way around. Also the language served as a model, a form, into which the students' knowledge was shaped. It indicated that there was something there to know. It also served as a standardized, stable tool for communicating even if the user didn't fully understand it. The vocabulary of the discipline and the workplace served as a tool into which knowledge had been distributed.

I have been treating conventions as stabilizing forces. However, because knowledge in this organization is necessarily unstable, the conventions making up genres are also never completely stable but only, as Catherine F. Schryer says, "stabilized for now" (1993, 204). An example of changing conventions also occurred in the exchange between Ben and Linda over her drawing. Among the instructions that Ben gave Linda was that placing numbers above the lines in a drawing (rather than at the end of the line) was "standard." This claim is a good example of how unstable knowledge can be in an organization. As I observed them, I had no doubt that he knew what he was talking about. However, when Linda later consulted another coworker to help her convert her drawing to the kind of dimensioning Ben asked for, this coworker told her that Ben was wrong. He said that the Pacific Equipment standard was for all text to be horizontal, not on the often slanting or vertical lines in a drawing. He said that this was an American National Standards Institute standard and that placing the dimensions above the line was from a Japanese International Standard that the engineering center had used for a while. But he said that now her group was the only one that used it. He declared it to be "nonstandard." This convention was one that had changed, although that change had evidently not yet reached all parts of the organization. This change shows both the provisional nature of conventional knowledge in an organization and the way in which changes in knowledge are linked to social structures. In the engineering center, this change in drawing conventions happened within the bounds of organizational groups. Newcomers learned the conventions of their groups, as what counted as knowledge varied from group to group. Indeed, their use of local conventions marked their group membership, so that the coworker who helped Linda conceded that she needed to use the "nonstandard" form because it was "standard" for the group in which she worked.

Thus, using models and conventions, including language conventions, seems to be one way that organizations deal with the instability of knowledge because imitation and replication are inherently conservative. Knowledge in a stabilized form can be distributed into these conventional forms so that everyone can access it with minimal effort. However, organizations necessarily exist within the tension between the need for sufficient stability to operate and sufficient flexibility to change, and even conventions change in order to reflect this learning environment. Because organizations are not monolithic, such changes may not occur simultaneously throughout the organization, and discontinuities can occur from department to department. The newcomers I observed valued learning these conventions because of the organizational knowledge they carried. Their coworkers too valued the access to knowledge that conventions of representation provided.

USING COMMUNICATION TOOLS
TO CREATE NEW KNOWLEDGE

The students, then, learned disciplinary and organizational language conventions as a means to access reified knowledge. However, if they were to be effective employees in the engineering center, they also needed to learn how to generate new knowledge. For the students, as for the regular employees, generating new knowledge was accomplished in large part through the effective use of communication tools. The term *communication* is sometimes taken to imply passing along information that already exists. However, when engineers at Pacific Equipment created texts of various kinds, they were often engaged in generating knowledge, not just passing it along. Additionally, creating knowledge is often taken to be a purely cognitive action, a product of pure thought, but at the engineering center, knowledge grew out of skilled interaction between people and the physical tools at their disposal.

For instance, one of the things that struck me as I watched the interns was the amount of effort they put into learning to use various software tools. Thus, when Sue indicated what she had learned during the summer, many of the things she mentioned were tools, and specifically tools such as Excel, Word, and Power Point, which were aimed at symbolic action. Learning to use these tools seemed important to the Pacific Equipment students' ability to accomplish engineering tasks. That is, the tools they used were not incidental to the communication and knowledge generation in which they were able to engage. Rather, knowledge about how to analyze and classify knowledge had become stabilized and accessible in the material form of the tool. Indeed, assuming that one had learned to identify a typified situation, the ability to generate a typified (i.e., generic) text was often spread over software and over the symbolic content one used it to generate. For instance, in vignette 5, we see Kevin creating his monthly report by entering text into the on-screen version of a previous month's report, so that software automatically formats his work to match what his supervisor will expect. As I pointed out in talking about work orders, repeated form is not the determining element in genre. As a matter-of-fact, when a repeated text looks like a collection of conventional language and format, such a perception is probably a sign that one is not (yet) an insider. However, if repeated form is expected, then newcomers should try to learn to generate it. Moreover, using repeated form can be a first step toward recognizing the typified situation to which a genre responds (cf. Winsor 1990b) and toward acting appropriately within that situation. For the engineering interns, cues about both repeated form and repeated situations were often embedded in software and the generation of knowledge in which the students were able to engage was facilitated by the software's use.

Adoption of a tool such as software is not a trivial matter in relations of knowledge/power. A person with a tool has different capacities and possibly different inclinations than a person without one. At Pacific Equipment, different people typically used different software. For instance, engineers were expected to give PowerPoint presentations; technicians were not. Nor were technicians expected to use software macros to analyze data. Access to these tools helped to enact the students' identity as engineers. As Gary Lee Downey puts it, humans and machines live inside one another. The machine is designed to allow humans to use it and is functional only to the extent that people are able to take in its function and workings. The human learns to behave as the machine or tool will allow. As Downey says:

> Think about each machine as a configuration of agencies—acts of positioning—that are part-human, and of each human as a configuration of agencies that are part machine. Life with this image might expect that adding a technology to an existing web of relationships would rearrange the identities of everyone and everything involved. . . . (1998, 6)

In school, we tend to think of learning as cognitive, as the mastery of concepts. Downey has described this idea of learning primarily as it exists among engineering students and faculty, who, he says "tended to understand learning as a cognitive process that takes place entirely inside the head" (1998, 134). It seems to me that the idea of learning he describes is widespread in academia. In contrast, Downey describes students learning to use Computer Aided Design systems in a way that involves disciplining their bodies as well as their minds. Computers and software programs (and writing) are physical entities and processes that one has to learn at least partly through physical effort. Bourdieu (1991) talks about disciplining the body as part of acquiring the "habitus," or disposition to act in certain ways, which marks people out for positions within a field. Using software for manipulating symbols was one of the bodily actions the engineering interns had to learn to undertake in order to be effective contributors to new knowledge.

All of the students were learning electronic tools, mostly on their own. The tools they were learning included the finer points of Word, PowerPoint, Outlook, and Excel, as well as an engineering drawing program called "Pro/Engineer." The tools also included electronic tools that coworkers provided, such as a macro that would automatically figure tolerances in a design. As is evident, these were primarily tools for communicating or for otherwise manipulating symbols and analyzing data. Using these tools marked the students as engineers because this symbolic work was character-istic of the engineering area of the facility and, indeed, is character-istic of all work aimed at generating new knowledge.

At the engineering facility, work tended to be geographically zoned. The fact that people were working in separate fields was fre-quently reflected in their working in separate physical areas. In the labs, work was aimed at manipulating vehicle and engine compo-nents to turn them into data. In contrast, in the engineers' cubicles, managers' offices, and meeting rooms, work tended to be conducted in various symbol systems ranging from natural spoken language to written texts of various kinds to computer programs to data analy-sis. This work was largely accomplished by means of various com-munication tools, including writing. As the photographs in chapter 3 show, the cubicles of the regular engineers were usually overflow-ing with files, notebooks, posted material, and paper of various kinds. Indeed one way to identify a student's cubicle was that it was usually far more barren because students had not had time to build up or accumulate corporate history and knowledge in the form of documents. Documents such as work orders and data reports allowed the lab to be linked with the cubicles and the cubicles to be linked to one another. Hutchins (1996) speculates that in a system of distributed cognition, communication between people may be one of the limiting factors in what can be accomplished because it is one of the limiting factors in the degree to which people are truly interacting. I believe it was the important role that these communication tools played in accomplishing engineering tasks, including the generation of knowledge, which led the students to spend so much time learning communication tools.

For instance, one of Sue's major tasks was a good example of the role that communication played at Pacific Equipment. The major part of her summer job was to serve as a condenser of data as they moved from the lab to the engineering area. She was the contact person for some of the tests being run on rival engines against which the company wanted to compare one of its engines. The company's intranet (a tool for communicating symbolic knowledge) had been set up so that the lab technicians could send her hourly data (taken by means of tools designed to represent knowledge symbolically) on whichever engine was being tested. The data went into a database program (a tool for analyzing symbolic knowledge) that she could use to check the information. The database, coupled with a software macro that her mentor gave her, allowed her to display the data in various graphs, which she found useful in interpreting the large array of numbers: "If it's not a straight line, you know . . . something [is] wrong," she said, demonstrating that she was thinking with her eyes here and not just with her head. Within the typified structure provided by these tools, Sue examined these data for what they might mean and then condensed them into a weekly report that she sent to the engineering group that wanted to compare its engine to the ones being tested. Many of the communication tools I saw her learning to use were related to this task. The tools were helpful because their structured nature made her work easier and also enabled her to proceed conventionally, thus making her work more useful to others. They enabled her to produce useful data reports, one of the most treasured genres in the engineering area.

Sue's data-transforming task was a common one in the engineering center, and it suggests that *communication* is too limited a term for the language and other symbolic activities in which she engaged. If we look at the way in which knowledge is generated and spread at Pacific Equipment, writing, speaking, and drawing can be seen as tools with which to analyze and generate knowledge, not just to pass it along. Sue's electronic manipulation of data is an example of "communication" tools being used for just such a purpose. When her report on the data was sent to the engineering group, it looked like she was passing information along, but by using the tools available to her, she had transformed the information before it left her hands. At the company, as information moved from the labs to the engineering area, it was typically transformed into more and more abstract forms used by (and useful to) people who were successively higher in the corporation.

In using tools to gradually move information from more concrete to more abstract forms, Sue's activity, and that of her fellow engineers, closely resembles the series of transformations that Bruno Latour (1999) claims is central to knowledge work. In a recent study, for instance, Latour described the work of a group of scientists gathering soil samples in the Amazon forest. As Latour watched these scientists work, he saw them executing dozens of steps using a tool of some sort to transform the data to a slightly more abstract state. The tools ranged from pieces of string used to divide the soil area into squares, to a set of boxes in which soil samples were categorized, to marks on charts and paper. The researchers moved from the concrete to the abstract by means of successive transformations using what Latour calls "hybrids," joint concrete/abstract devices such as data tables or categorizing storage units. People weren't able to learn anything through direct cognitive contact with raw nature. They are not what Latour calls a "brain-in-a-vat" (4), a mind working apart from a physical environment. Tools, and particularly tools associated with "communication," enable knowledge to be built up through gradual, shared, physical/mental activity. Cognition has been distributed among different people and also between people and their tools.

The texts that the students produced resembled the hybrids Latour (1999) describes. They used institutionalized, predetermined forms for structuring abstract information. The software tools they used provided them with easier access to these forms. The generic texts the students thus generated provide typified responses to typified situations. Indeed, genre is always a combination of concrete and abstract elements of the kind Latour refers to as hybrid.

Thus it is not surprising that the students in this study spent so much time learning communication tools. What look like "communication" tools—that is, tools for passing along already created knowledge—are also knowledge-organizing, knowledge-creating tools that help the students to do engineering work. So, for instance, in Sue's work, the lab technicians had already used tools to convert the engine into abstract numbers on a computer screen or on paper. Sue moved that transformation along another step. Don had a task that was similar to Sue's. He entered data from the labs into a spreadsheet. Then he had a macro (provided by his mentor) that allowed him to graph the data using various parameters, such as "NOx versus Mode" or "PM versus Mode." He told me that these parameters were the ones his mentor usually graphed, so he assumed that they reflected the most important comparisons and had taken them as a model for his own work. I watched him flash these graphs up on his screen very briefly. Their visual nature allowed him to see how the engine was doing and where problems lay very quickly. For instance, one graph had an area outlined in red in its lower left-hand corner. All of the data fell outside this area, an unfortunate situation because ideally they would all be inside the area. This writing created knowledge. Don wrote it (although the instruments that collected the data, the computer program that created the graphs, and the mentor who created the computer program also wrote it). But Don didn't know ahead of time what it had to say to him. He used the graphs to decide where to set his parameters for the next round of tests. Then he used the graphs to communicate the information to his coworkers. Whether writing was knowledge creating or knowledge communicating depended on which direction you were looking, upstream to what existed before it or downstream to what would be done next. As Latour says, "Each stage is matter for what follows and form for what precedes it" (1999, 74).

So communication tools help to produce small bits of new knowledge. The students in this study drew on the distributed cognition structured into communication tools and were thereby enabled to become part of overall system of knowledge creation at the engineering center. Organizational genres may frequently function (among other things) to help users generate needed knowledge. Even these interns were able to participate in creating knowledge by using tools and the support of others, that is, by becoming connected to distributed cognition. In so doing, they exercised the power that comes from shaping the meaning they found in their workplace, although they did so in rather minor ways. The shaping of the workplace and their identity was reciprocal.

FINDING THEIR PLACE IN THE HIERARCHY

When these interns accessed the distributed cognition at the engineering center, they positioned themselves socially as engineers. This positioning gave them a place in Pacific Equipment's hierarchy as well as in its knowledge generation. For instance, their relationship with the lab technicians was an interesting one. In general, the students I observed got along reasonably well with the technicians and sometimes acknowledged that the technicians knew more than they did. Sue told me that the technicians were "very helpful," and Don told me that when something went wrong with the tests for which he was responsible, the technicians were the people he first asked for help. Moreover, the technicians sometimes seemed to engage in teaching the interns in much the same way that the engineers did. For instance, while Don and I were waiting for the technician who worked with him, another technician in the same area explained a test he was running. Don was not involved in this testing, so the technician told him about it simply because he believed Don didn't know about it (which in fact was true) and he thought it was worth knowing (which again Don confirmed).

However, despite the fact that they often knew less than the technicians did, the students' positioning as novice engineers placed them in authority over the technicians, a fact that sometimes resulted in what I thought were odd situations. For instance, Sue told me that when she wanted to write a work order instructing the technicians to perform a certain task, "a lot of times, I'd just go out to the mechanic and say I need this done. Tell me how to say it. And the mechanic would write it down for me." However, Sue had to sign the work order because, as an engineer, she was in the position of authorizing the work, and power came from her position relative to the technicians rather than from her personal qualities.

Sue's experience in regard to work orders was not unique at the engineering center. Although work orders always originated in the engineering area, they were not always written by engineers. While I was at Pacific Equipment, one engineering area used a co-op student to write some of its work orders and another used a retired technician who had been hired back in the engineering area to serve as a liaison with the lab. This variability in writers suggests the degree to which work orders' authority came from the social roles the engineering and lab areas played in the organization, rather than from the education or personal authority of the writer. As Bourdieu says, a "spokesperson is only able to use words to act on other agents and, through their action, on things themselves, because his speech concentrates within it the accumulated symbolic capital of the group which has delegated him and of which he is the *authorized representative*" (1991, 111, emphasis in original).

In contrast to the plentiful interaction the students had with those below engineers in the hierarchy, they saw very little of those more than one level above them. Their status as interns usually meant that they were not directly responsible for their work to managers outside their group; rather their mentor or supervisor was usually held responsible for making sure their work was usable.[2] They were probably most likely to see managers when they gave presentations that were required from all interns at the end of the summer. Don said that the audience for these presentations would typically include other technical people, managers, and people from the Human Resources (HR) Department. Don had not done such a presentation before and was uncertain whether it should be about his technical project or about what he had done and learned over the summer, confusion that seemed to be natural in the face of his mixed audience and in the face of his dual role as employee and student. HR people, for instance, or even managers from other areas, were unlikely to understand his work on an EGR valve in any detail, but it's very likely that they wanted to be able to assess him for further employment. Don's supervisor eventually advised him to make his presentation as technical as possible, saying that "it is more beneficial to you professionally to make it technical, not [to say] what you did on your summer vacation." In other words, the supervisor believed that Don's "presentation" was of himself as an engineer rather than as a student on summer vacation. Such a stance would presumably make it seem more desirable to hire Don as an engineer once he had graduated. If Don's experience was typical (and my observations of the other students suggest that it was), in what is their most intense meeting with managers during the summer, students were encouraged to display the changes in their identities, to sound like engineers, in order to convince managers that, like an engineer I spoke of in chapter 3, they "knew what they were talking about." Such a display is not a false front, but rather a demonstration of the extent to which they have acquired the discursive tools of engineering and thus will be more likely to contribute to its distributed cognition.

BECOMING AN ENGINEER:
A MATTER OF KNOWLEDGE AND POSITION

When newcomers enter the engineering center, then, they enter a system of distributed cognition that functions and is held together at least partly by means of communication tools. For the Pacific Equipment interns, learning involved getting plugged in to the activity system around them. Learning also meant spending a great deal of time learning about tools, and particularly communication tools. These communication tools ranged from software to vocabulary to genres. As I have demonstrated, the newcomers spent much time learning these communication tools because they were the essential means by which the system functioned and thus were the essential means by which the newcomers could be plugged into the network (cf. Freedman and Smart 1997). This study also suggests that more established employees also engaged in learning as the workplace changed on a daily basis; indeed "changed" might be taken as a synonym for "learned." As Jean Lave has said,

> Situated activity always involves changes in knowledge and action . . . and "changes in knowledge and action" are central to what we mean by "learning." . . . People in activity are skillful at, and are more often than not

engaged in, helping each other to participate in changing ways in a changing world. . . . [P]articipation in everyday life may be thought of as a process of changing under-standing in practice, that is, as learning. (1995, 5–6)

Plugging into this system is a difficult task because it is in con-stant flux, so that even more experienced employees spend a great deal of time learning. Such learning is accomplished not just by mental activity, but also by physical and social experience. We cannot think of learning and knowledge as isolated from goals, col-leagues, history, and material reality. Learning is not separate from the rest of activity. It is the means by which we remain attuned to activity and able to function within it.

Language practices such as vocabulary or genres that were con-ventional in engineering in general and at Pacific Equipment in particular gave the students and regular employees access to previ-ously stabilized knowledge that had been distributed to them. For the students, learning conventional language also allowed them to reshape their identity as engineers and to gain access to distributed cognition. In turn, their identity as engineers placed them in the organizational hierarchy and authorized them to speak and write in certain ways. Through these authorized language practices, they were empowered to contribute to the company's engineering effort.

The newcomers were able to generate new knowledge by using communication tools and the distributed cognition that was embed-ded in that use. Communicating existing knowledge is only part, albeit an important part, of what writing at work achieves for dis-tributed cognition at work. Communication tools are also a central means by which groups jointly construct new knowledge. Even as newcomers, the students were enabled to help construct new knowl-edge by using communication tools to analyze data in however small a way and then passing the analyzed data along to whoever would carry out the next step in the distributed process of knowledge con-struction. Assuming the social role of engineer also entitled and enabled them to assume a role in generating engineering knowl-edge. Thus, learning language practices was essential to entering this system of knowledge/power. Learning engineering genres pro-vided them both with support for the tasks they needed to accom-plish and an appropriate engineering voice.

CHAPTER 6

KNOWLEDGE/POWER/TEXTS
IN AN ENGINEERING CENTER

I now want to return to the research questions I asked in chapter 1 and to try to summarize the answers to them that this study of Pacific Equipment has suggested:

- How do texts, and particularly generic texts, help technicians, engineers, and managers to generate the cultural capital of knowledge?
- How is the textual generation of various kinds of capital used to create and occupy positions of power? What does this use of texts tell us about genre?
- How does power affect the generation of text?

In particular, I want to examine some of the implications this study has for our understanding of the complex, reciprocal relationship among knowledge, power, and text. I will also articulate some of the implications this study has for organizations that hire engineers (and other specialists) and for universities that educate them.

GENERIC TEXTS ENABLE DISTRIBUTED COGNITION

One of the first conclusions we can reach about generic texts is that they enable distributed cognition and thus help to generate knowledge. At Pacific Equipment, as elsewhere, texts respond to recurring social situations by taking on recurring purposes and forms that we call "genres." These genres are enacted in individual use but are shaped by many forces including disciplinary and local conventions, such as those governing graphs or notation forms on engineering drawings, tools such as software, and regulated formats required by the organization. The very existence of such texts, along with their commonly recognized forms, facilitates the distribution of cognition among technicians, who generate data, engineers who analyze it, and managers who set the agenda. People in an engineering center, and presumably in many other kinds of workplaces, can divide the labor of generating knowledge and coordinate that divided labor by means of texts. Texts allow knowledge to be reified and preserve it across time and distances that range from the space between cubicles to the space between continents. Because texts have to be mutually acceptable to a number of people, their creation also serves as an occasion upon which commonly accepted knowledge can be negotiated. For instance, when engineers agree on the interpretation of a data curve or work together to create a drawing, the joint interpretation or creation of these texts is the event by which they negotiate their understanding. Moreover, when such texts are generic, their recurring form guides people so that agreement is made easier. Thus texts and the act of creating them are part of the social fiber that knits people together and that enables them to undertake joint enterprises, including the generation of knowledge.

At the engineering center, this particular impact of generic texts was especially valuable for summer interns. Using conventional language practices allowed the interns to have their attention drawn to what was important in the previous work done at the engineering center. They could imitate previous reports or drawings and in so doing "leverage" the experience of past employees rather than start from scratch in their efforts to function as engineers. Although this advantage was particularly evident in the interns' actions, it was actually present for everyone. The generic nature of texts made it easier for a wide variety of people to access existing knowledge because, in these genres, the knowledge was reified and thus easy to recognize and understand. People in the engineering center had distributed part of their knowledge and practice into their conventional language enabling others to use it in their part of the distributed cognition in effect at Pacific Equipment.

In addition to easing access to existing knowledge, generic texts also can serve as tools to develop and shape new knowledge and then to reshape knowledge into forms that are useful to different parts of the organization. Within the engineering center, texts allow for the successive processing of data into engineering knowledge and then the conversion of that engineering knowledge into monetary equivalencies by which the corporation acts. For instance, laboratory reports consisting of numbers were sent to one of the interns, Sue, who used a computer macro that her mentor had provided in order to structure and make sense of the numbers. The graphs she created with the macro's help were then sent along to an engineer who used that information to judge how Pacific Equipment's product was faring against a rival product. Eventually, the engineer would write a report about his judgments and that information, along with other streams of information that his group had been processing, would be sent to a supervisor, who would prepare a budget request to finance further engineering work. At each step of the way, various kinds of texts were used to construct knowledge that someone would find useful. Specific genres were useful for people in specific organizational roles.

Thus the creation of useful texts was made easier by the existence of genres that met commonly occurring situations within the corporation. The use of these genres was promoted by the existence of software that embedded recurring questions and also by the existence of documentation that superiors required subordinates to prepare repeatedly. Generic tools such as software and documentation formats directed writers' attention to what their superiors wanted to know and supported their creation of appropriate answers. That is, the use of some genres did not occur spontaneously. Rather, managers and coworkers required them within the coordinated system of distributed cognition in play at Pacific Equipment. These requirements served as centripetal forces (Bakhtin 1981) in maintaining the structure and role of institutionalized genres, despite the tug on these forms that was exerted as workers in subordinate fields tried to maintain control of their own work. As is suggested in figure 6.1, such institutionalized genre use is an important tool linking the creation of knowledge and the creation of power.

Fig. 6.1. *Forces Shaping Institutionalized Genres*

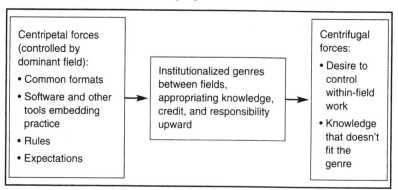

TEXTS ARE A MEANS TO CREATE
AND OCCUPY POSITIONS OF POWER

Another conclusion we can draw about generic texts based on this study is that they are a means that is used to create and occupy positions of power. Texts function as immutable mobiles to allow the accumulation and appropriation of various kinds of capital in centralized hands. That is, the results of distributed cognition can be represented as belonging to the organization as a whole but also can be credited to some people more than to others because of the way in which texts represent the knowledge and the work behind it. For instance, instrument traces allow the knowledge generated by the laboratory technicians to be accumulated in the hands of the engineers. Work orders allow the engineers to be credited with what happened in the lab even though engineers did not actually physically perform the laboratory work. Similarly, reports allow engineers' knowledge to be accumulated in the hands of managers.

In addition to allowing the processing of data into knowledge, texts also allow the processing of laboratory and engineering knowledge into profit-making activity for the corporation. One of the primary texts used to accomplish this is the budget, which translates knowledge into the terms of monetary capital and translates monetary capital into a resource for the generation of engineering knowledge while keeping control in the hands of managers. By controlling the budgeting of money, managers can also control the time that engineers and technicians will spend on various activities and thus shape the work in the engineering center.

Thus texts help to create powerful positions from which designated people can operate. From these positions, certain people are allowed to use certain texts. They can write in certain ways because they are authorized to do so. For instance, people in the engineering area are authorized to write work orders by means of the positions they hold, even when they are students. Using this authorized language also helps to position people as engineers, or technicians, or managers. Within any organization, commonly recognized genres are most likely to be made up of this authorized language. The very institutionalization that leads to its common use is most likely to enable us to perceive such language as generic. Such institutionalized genres are one of the primary means for distributing knowledge/power to the advantage of those in control.

POWER RELATIONS BETWEEN WRITERS AND READERS AFFECT GENRE CHOICE

Rhetoricians now define genre as a typified textual reaction to a typified social situation. Within hierarchical organizations, part of that typified situation is inevitably the relative positioning of readers and writers. For instance, at Pacific Equipment, typical engineering texts seem to be of three kinds: work orders that were sent to technicians, data texts such as instrument traces that they used to negotiate common knowledge among themselves, and more sustained pieces of prose that were sent as reports to managers. Work orders reflected the authorized organizational belief that engineers controlled the work in the lab, although everyone involved also knew that technicians brought expertise and initiative to laboratory work. The structure of work orders reflected the relative positions of engineers and technicians within the corporate hierarchy more accurately than it did the actual conduct of work. That is, the power relations between engineers and technicians were an extremely influential part of the recurring situation to which the work orders had to respond.

The fairly harmonious negotiation around data texts within engineering suggests that engineers agreed upon the symbolic capital they accorded to this kind of evidence. Thus they did not see such texts as persuasive or implicated in power struggles. In contrast, when they wrote texts that would cross the gap from the field of engineering into the larger field of the for-profit corporation, they knew that goals and standards for a "good" product were somewhat different, and they believed they had to create texts that would persuade managers to allow them to do their work as they saw best. They had to convince managers that they "knew what they were doing." The moment when writers address an audience outside of their own field is often one when power relations surface most obviously. At that moment, writers must often shift from using genres that are shaped by the disciplinary training and local practices they share with colleagues to using genres that will serve the purposes of the dominant field rather than their own. At Pacific Equipment, this moment was also the one in which writers became most conscious of the need to be rhetorically effective if they wished to maintain control over their own work.

Thus the recurring social situation that Carolyn R. Miller (1984) has taught us to see as essential to genre must include the power relationships between readers and writers and the degree to which they share common knowledge and concerns. The power relations in effect in an organization frequently lead to the institutionalization of genres meant to facilitate the work of the dominant group. These genres serve to regulate the exchange of information and claims to knowledge among fields in ways the dominant group has deemed appropriate. As JoAnne Yates and Wanda Orlikowski point out, "A genre established within a particular community serves as an institutionalized template for social interaction" (2002, 15). This social interaction includes the distribution of power. Genre systems predate the arrival of most employees in organizations. They structure the work lives of these employees and are, in turn, used by the employees to structure the work of others.

IMPLICATIONS FOR ORGANIZATIONS AND UNIVERSITIES

This study also has a number of implications for organizations that employ engineers (or other specialists) and for the universities that educate them. If an organization that depends on distributed cognition is made up of more than one field, managers need to make certain that people in different fields more or less share the overall goal of the organization. This study shows Pacific Equipment managers doing that by a variety of textual means, including more coercive texts such as budgets and more motivationally oriented texts such as mission statements. But managers need to remember that people in the various fields will also have knowledge-related goals of their own with which managers may not be acquainted. This study suggests that the organization's distributed cognition is likely to work better if these people's goals are also met as much as possible. Using slightly different terms, John Seely Brown and Paul Duguid argue that companies are not unified monoliths, but instead consist of multiple communities of practice. They further assert that if people in positions of power attempt to overcontrol these communities of practice (what I am calling by Bourdieu's term, *fields*), they can stifle the generation of knowledge:

> Alternative worldviews . . . are inevitably distributed throughout all the different communities that make up the organization. For it is the organization's communities, at all levels, who are in contact with the environment and involved in interpretive sense making, congruence finding, and adapting. It is from any site of such interactions that new insights can be coproduced. If an organizational core overlooks or curtails the enacting in its midst by ignoring or disrupting its communities-of-practice, it threatens its own survival in two ways. It will not only threaten to destroy the very working and learning practices by which it, knowingly or unknowingly, survives. It will also cut itself off from a major source of potential innovation that inevitably arises in the course of that working and learning. (1996, 76)

Thus people in more powerful positions need to resist the temptation to institutionalize only those genres that forward their own field's goals. For instance, engineers at Pacific Equipment probably need to make it officially possible for the insights of laboratory technicians to be recorded and circulated. To maintain the current one-way system of work orders is to squander a resource that might help lead to the generation of knowledge.

Engineering schools also need to consider the negotiation of knowledge/power between fields to a larger extent than they currently do. As is probably typical of professional schools, engineering schools are primarily aimed at preparing students to generate knowledge within their disciplinary field. Students learn to generate the data-displaying texts that are most valued in engineering and then, to a lesser degree, to explain how these data support design decisions they have made. The audience for these texts is other engineers who share many of same values and assumptions. Some design courses ask students to consider costs but students have little experience in negotiating with budget-conscious superiors. Perhaps most importantly, they are not trained to see such negotiation as a normal part of engineering. As a consequence, young engineers are often surprised and annoyed by managers' "interference" in their work. Young engineers (and sometimes more established ones) need to learn to engage in such negotiation in a way that recognizes the manager's position as legitimate at the same time that it energetically presents the position of the engineers. I'm not sure if such skills can be taught in school because no matter how much the engineering professor tries to simulate a corporate atmosphere, students tend to respond to the underlying reality of the school setting (Freedman, Adam, and Smart 1994). Indeed, the interns who worked at Pacific Equipment suggested that they had learned even the engineering skills they needed during their internships rather than during their academic preparation, so the extent to which students can learn corporate negotiation skills in school is doubtful. However, engineering and communication professors can at least make students aware that such negotiation is normal. And companies that hire interns can make the need for such negotiation explicit.

KNOWLEDGE/POWER/TEXT

What we see at Pacific Equipment, then, is that knowledge, texts, and power are all deeply intertwined. It is hard to imagine how any one of them might function without drawing upon and leading to the presence of the other two. Our contemporary scholarly examinations of the relation between rhetoric and power have tended to move directly into questions of ethics. And in those examinations of ethics, we have tended to be deeply suspicious of power, to see power solely as dangerous and open to abuse. Thus, for instance, we caution researchers to set limits on their own power over study participants (Newkirk 1996), and we worry that writing can be so effective that it can empower the writer to trample on the rights of others (Tillery 2001). I share these concerns because it is obvious to me that writing "well" means writing powerfully. In Pierre Bourdieu's (1991) terms, I know that I can use writing to generate a powerful position for myself in my own field, and I try to be careful to use that power well.

However, I also believe that in treating power solely as a question of an individual's ethics, we are treating it as a heroically achieved, personal possession (cf. Horner in press). But power is always relational and is always achieved within preexisting social structures in which we learn to use various means to achieve desirable positions. I can use texts to create a powerful position for myself within rhetoric because the field exists and makes certain discursively structured power positions available to me. I have to work within the institutionalized genres of that field to make its other members read my knowledge/power struggle as a question of knowledge. This is how I am able to move into a position of power. Occupying a powerful position is neither a sign of individual heroism nor an automatic sign of corruption on my part.

We might be tempted to judge the system I have described at Pacific Equipment as a pernicious one. However, I do not believe that all exercises of power are pernicious. Power is a way to make things happen. It can be productive. But in hierarchical companies like Pacific Equipment, power is never evenly distributed. The uneven distribution of power is not due to individual heroic accomplishments that result in merited differences. Rather it is accomplished in the systemic use of sociotechnical means, including generic texts such as work orders that ordinarily slip unnoticed beneath the surface of everyday life. This study suggests that power is constructed in the trivialities of everyday life that are so taken for granted as to be transparent to us. If we wait for major events to come along, we will miss how ordinary life happens. We will not be able to see how the technicians' work becomes folded into the engineers' work and how the engineers' work becomes folded into that of management so that in the end what we have is a product credited to Pacific Equipment and not to any individual employee. We will also fail to see how managers are given greater credit and reward for the product, as they are also given greater responsibility and blame when things fail. These conditions will seem natural to us, rather than the accomplishments they are.

I certainly do not want to argue that power is harmless and innocent, but I do want to argue that it is unavoidable and potentially productive. I do not seek to be powerless and I do not wish powerlessness on my research participants and students. What I do seek is to understand how power, knowledge, and text can generate one another so that I can be conscious of what I am doing. I also seek to make others conscious of the consequences of their own actions and of the actions of those for whom they work or with whom they otherwise associate (see Tillery 2001 for a discussion of this point). If we see power as automatically suspect, we will tend to be blind to the ways in which our own actions lead us into powerful positions. Moreover, we will not understand how others generate the powerful positions they hold. I hope that this examination of the Pacific Equipment engineering center will move us toward a more sophisticated understanding of knowledge/power/text.

NOTES

CHAPTER 1. USING WRITING TO
NEGOTIATE KNOWLEDGE AND POWER

1. I gathered information for this book in the five summers from 1996 through 2000, by shadowing eighteen people in the engineering center for several different two- to four-hour stretches and then interviewing them. In total, I spent over a hundred hours observing participants and additional time interviewing them. I varied the days of the week and times of the day I observed them because workweeks often have a pattern that varies from day to day.

As I observed these people, I took handwritten field notes as to what they did, whom they interacted with, what communication they engaged in, and so on. I took notes by hand rather than on a laptop because my participants moved around the facility a great deal and I needed to be able to follow them and to take notes while walking. This mobility was a sign of their need to interact with others in order to do their work, that is, a sign of distributed cognition. I transcribed field notes within twenty-four hours of each observation, adding analytic notes containing comparisons to other observations I had made, comments on what I found surprising, and questions and preliminary interpretations that I explored in further observations and in interviews. (See Doheny-Farina and Odell 1985 for an explanation of types of field notes. See Glaser and Strauss 1967 for a discussion of the process of cycling between data and theory resulting in what they call "grounded theory.") This process resulted in 209 pages of single-spaced typed field notes.

I conducted audiotaped interviews with most of the participants, asking questions that I developed from my observations.
Ken's schedule did not allow me to interview him, nor was I able to
formally interview the lab technicians, whose time was regulated.
However, the technicians were usually able to answer my questions
as they worked and were quite willing to do so. Two engineers asked
that I not tape their interviews, although they did not mind if I took
notes which I later typed up. I fully transcribed all the tapes I had.
Interview transcripts and notes amounted to 133 single-spaced
typed pages. I also collected samples of documents participants were
creating or using while I was there. These included reports, letters,
sketches, work orders, E-mail, a student self-evaluation, and pages
from a manual.

As my description of information gathering implies, I engaged
in ongoing preliminary analysis as I transcribed my field notes. Following common practice in observation research (Strauss and
Corbin 1990, 197-223; Miles and Huberman 1994, 72–76; Bishop
1999, 79-80), I also engaged in ongoing analysis by writing memos
about themes that recurred from observation to observation. I analyzed this material from several different perspectives as I prepared
articles and presentations. In writing this book, I reframed the
results of my analysis to consider the interrelationship of power,
knowledge, and text. I shared drafts of my work with the participants and asked them to tell me if I was saying anything that was
flat out wrong or anything that would be harmful to Pacific Equipment or to them personally. No one objected to my conclusions and
several people said they agreed with my depiction of the company.
Doug said that it was a "very accurate" picture of how the company
operated.

2. I use pseudonyms for all of the participants and for the company itself. I have also changed some minor details in order to pro-
tect the privacy of those involved. My research was approved by the
Institutional Review Board for Human Subjects Research at Iowa
State University.

3. Using the term *social capital* in a slightly different way than
Pierre Bourdieu does, management theorists have examined the
question of how social good can accrue to individual people or to

social units as they use their relative status in a group (Unseem and Karabel 1986; Burt 1997) or to the group itself as a result of the collective efforts of its members (Putnam 1993, Fukuyama 1995). Carrie R. Leana and Harry J. Van Buren (1999) argue that social good must be shared by communities and individuals within those communities if it is to be sustained. Their argument has implications for how a company like Pacific Equipment supports its employees and draws support from them.

4. Managerial literature sometimes differentiates between power, authority, and influence. See Susan M. Katz (1998, 420–422) for a discussion. Here, I use "authority" to mean institutional recognition of one's right to control one's own actions or those of others. I use "power" to mean the ability to exercise that control. I treat "influence" as one sign of power.

5. It is, of course, possible to define a "successful corporation" in a way that includes qualities other than profitability. Ben and Jerry's, for instance, gave social responsibility a prominent place in its corporate practice.

VIGNETTE 2. TWO HOURS IN AN AFTERNOON OF A MANAGER: DOUG

1. I do not examine text's move on-line in this book, although it was clearly important. The use of E-mail increased greatly during the five years I was an observer at Pacific Equipment. As I note in vignette 1, a contractor who was working on lab renovations told me that I could not possibly understand what was going on at Pacific Equipment unless I had access to E-mail. I did read E-mail over people's shoulders, but the increasingly virtual nature of communication was not my focus here. However, I do wonder if the movement of communication on-line affects its visibility and thus its openness to scrutiny. E-mail looked less visible than paper text to me, but Cheryl Geisler (2001) argues in the opposite direction that text's move on-line makes it more publicly available to those with shared technology, an argument I find provisionally convincing. See also Geisler and her colleagues (2001).

CHAPTER 2. MANAGING THE ORGANIZATION THROUGH POWERFUL TEXTS

1. The need to separate the design and testing groups exemplifies a point that Edwin Hutchins makes about distributed cognition. When a group is trying to choose among several interpretations of experience, some separation of group members is desirable because it allows them to avoid "confirmation bias" (1996, 239) or the tendency to hold to already accepted interpretations and to disregard negative evidence. Separated members or groups will be more likely to develop separate interpretations. In other words, in thinking about complicated questions, there is such a thing as too much communication. As Hutchins says, "All the strategies that overcome confirmation bias work by breaking up continuous high-bandwidth communication" (262).

2. Brad had come to Pacific Equipment from the marines. After the meeting with Ken, he told me that in his previous workplace, there would never have been such a "touch, feely" meeting. The commanding officer would have declared Brad to be "the advanced technology guy" and would tell everyone else to support him. Assuming that Brad is right and that such a procedure would work, the degree to which managers can order change must vary from organization to organization. That is, it relies on an institutional organization of powerful positions rather than on power that any one individual "has." The military obviously organized powerful positions differently than the engineering center is able to do.

VIGNETTE 3. A MEETING WITH ENGINEERS: JOHN

1. Note that the theme of meeting customers' needs runs throughout this meeting. Engineers at this meeting seem to be trying to enlist customers' needs as an argument for the funding of their project, probably because as an argument, it combines the engineering value of quality and the management value of potential profitability. Thus it can function as an argument that engineers and managers both accept as valid.

CHAPTER 4. AMASSING KNOWLEDGE IN THE
HANDS OF THE MORE POWERFUL

1. Interestingly, differences in income were not clear markers of the relative statuses of the technicians and engineers. In 1997, the technicians' base salary was approximately forty seven thousand dollars per year, plus benefits and overtime. In many ways, Pacific Equipment valued these workers.

2. One technician told me that the account number was the most important piece of information on the work order because it meant that he would be paid. Thus we see again the way in which forms can be used to create an equivalence between different kinds of capital. In this case, work and the knowledge it generates are converted into monetary terms.

CHAPTER 5. ENTERING SYSTEMS
OF KNOWLEDGE/POWER

1. See Michael Cole (1996) for a discussion of what constitutes a tool. In general, Cole claims that tools have both a material and conceptual manifestation. The material manifestation of language would be in such forms as brain wave, ink, or sound.

2. Regular engineers also have their work approved by others. One engineer with twenty-two years of experience at Pacific Equipment told me that for the first ten years or so, engineers have an "approver" who "blesses" their work. He still has such an authority to whom he answers but can use his own judgment on when he needs to get some conclusion or report approved. He does so only when he is uncertain of his own work. Thus the experience of interns can be seen as on a continuum with that of regular engineers.

BIBLIOGRAPHY

Amann, K., and Karin Knorr-Cetina. 1990. The fixation of (visual) evidence. In *Representation in scientific practice*, eds. Michael Lynch and Steve Woolgar, 85–121. Cambridge: Massachusetts Institute of Technology Press.

Amerine, R., and J. Bilmes. 1990. Following instructions. In *Representation in scientific practice*, eds. Michael Lynch and Steve Woolgar, 323–335. Cambridge: Massachusetts Institute of Technology Press.

Artemeva, Natasha, and Aviva Freedman. 2001. 'Just the boys playing on computers': An activity theory analysis of differences in the cultures of two engineering firms. *Journal of Business and Technical Communication* 15 (2): 164–194.

Bakhtin, Mikhail. 1981. *The dialogic imagination*, ed. Michael Holquist. Austin: University of Texas Press.

Barley, Stephen R., and Julian E. Orr. 1997. Introduction: The neglected workforce. In *Between craft and science: Technical work in U.S. settings*, eds. Stephen R. Barley and Julian E. Orr, 1–19. Ithaca: Cornell University Press.

Bazerman, Charles. 1997. Discursively structured activities. *Mind, Culture, and Activity* 4 (4): 296–308.

Beaufort, Anne. 1999. *Writing in the real world: Making the transition from school to work*. New York: Teachers College Press.

Berkenkotter, Carol, and Thomas N. Huckin. 1995. *Genre knowledge in disciplinary communication: Cognition/culture/power*. Hillsdale, NJ: Erlbaum.

Bishop, Wendy. 1999. *Ethnographic writing research: Writing it down, writing it up, and reading it.* Portsmouth, NH: Heinemann.

Blakeslee, Ann. 1997. Activity, context, interaction, and authority: Learning to write scientific papers in situ. *Journal of Business and Technical Communication* 11 (2): 125–169.

———. 2000. *Interacting with audiences: Social influences on the production of scientific writing.* Mahway, NJ: Erlbaum.

Bourdieu, Pierre. 1991. *Language and symbolic power.* Translated by Gino Raymond and Matthew Adamson. Edited by John B. Thompson. Cambridge: Harvard University Press.

———. 1993. *The field of cultural production: Essays on art and literature.* Edited by Randal Johnson. New York: Columbia University Press.

Brown, John Seely, and Paul Duguid. 1996. Organizational learning and communities-of-practice: Toward a unified view of working, learning, and innovation. In *Organizational learning*, eds. Michael D. Cohen and Lee S. Sproull, 58–82. Thousand Oaks, CA: Sage.

Bucciarelli, Louis L. 1988. Engineering design process. In *Making time: Ethnographies of high-technology organizations*, ed. Frank A. Dubinskas, 92–122. Philadelphia: Temple University Press.

———. 1994. *Designing engineers.* Cambridge: Massachusetts Institute of Technology Press.

Burt, Ronald S. 1997. The contingent value of social capital. *Administrative Science Quarterly* 42 (2): 339–365.

Chin, Elaine. 1994. Redefining 'context' in research on writing. *Written Communication* 11 (4): 445–482.

Cohen, Michael D., and Lee S. Sproull, eds. 1996. *Organizational learning.* Thousand Oaks, CA: Sage.

Cole, Michael. 1996. *Cultural psychology: A once and future discipline.* Cambridge: Belknap.

Cole, Michael, and Yrjo Engstrom. 1993. A cultural-historical approach to distributed cognition. In *Distributed cognitions: Psychological and educational considerations*, ed. Gavriel Salomon, 1–46. New York: Cambridge University Press.

Cross, Geoffrey A. 1994. *Collaboration and conflict: A contextual exploration of group writing and positive emphasis.* Creskill, NJ: Hampton.

Dias, Patrick, Aviva Freedman, Peter Medway, and Anthony Pare. 1999. *Worlds apart: Acting and writing in academic and workplace contexts*. Mahwah, NJ: Erlbaum.

Dias, Patrick, and Anthony Pare. 2000. *Transitions: Writing in academic and workplace settings*. Cresskill, NJ: Hampton.

Doheny-Farina, Stephen, and Lee Odell. 1985. Ethnographic research on writing: Assumptions and methodology. In *Writing in nonacademic settings*, eds. Lee Odell and Dixie Goswami, 503–535. New York: Guilford.

Downey, Gary Lee. 1988. *The machine in me: An anthropologist sits among computer engineers*. New York: Routledge.

Faber, Brenton. 2002. *Community action and organizational change: Image, narrative, identity*. Carbondale: Southern Illinois University.

Fairclough, Norman. 1989. *Language and power*. London: Longman.

Foucault, Michel. 1979. *Discipline and punish: The birth of the prison*. Translated by A. Sheridan. New York: Random.

Foucault, Michel. 1980. *Power/knowledge: Selected interviews and other writings 1972–1977*. Translated by Colin Gordon, Leo Marshall, John Mepham, and Kate Soper. Edited by Colin Gordon. New York: Pantheon.

Freedman, Aviva, and Christine Adam. 1996. Learning to write professionally: 'Situated learning' and the transition from university to professional discourse. *Journal of Business and Technical Communication* 10 (4): 395–427.

Freedman, Aviva, Christine Adam, and Graham Smart. 1994. Wearing suits to class: Simulating genres and simulations as genre. *Written Communication* 11 (4): 193–226.

Freedman, Aviva, and Peter Medway. 1994. Locating genre studies: Antecedents and prospects. In *Genre and the new rhetoric*, eds. Aviva Freedman and Peter Medway, 1–20. Bristol, PA: Taylor & Francis.

Freedman, Aviva, and Graham Smart. 1997. Navigating the current of economic policy: Written genres and the distribution of cognitive work at a financial institution. *Mind, Culture, and Activity* 4 (4): 238–255.

Fukuyama, Francis. 1995. *Trust: The social virtues and the creation of prosperity*. New York: Free Press.

Geertz, Clifford. [1983] 2000. *Local Knowledge*. New York: Basic.

Geisler, Cheryl. 2001. Textual objects: Accounting for the role of texts in the everyday life of complex organizations. *Written Communication* 18 (3): 296–325.

Geisler, Cheryl, Charles Bazerman, Stephen Doheny-Farina, Laura Gurak, Christina Haas, Johndan Johnson-Eilola, David S. Kaufer, Andrea Lunsford, Carolyn R. Miller, Dorothy Winsor, and Joanne Yates. 2001. IText: Future directions for research on the relationship between information technology and writing. *Journal of Business and Technical Communication* 15 (3): 269–308.

Giddens, Anthony. 1984. *The constitution of society: Outline of the theory of structuration*. Berkeley: University of California Press.

Glaser, Barney G., and Anselm L. Strauss. 1967. *The discovery of grounded theory: Strategies for qualitative research*. Chicago: Aldine.

Hansen, Craig J. 1995. Writing the project team: Authority and intertextuality in a corporate setting. *Journal of Business Communication* 32 (2): 103–122.

Henderson, Kathryn. 1999. *On line and on paper: Visual representations, visual culture, and computer graphics in design engineering*. Cambridge: Massachusetts Institute of Technology Press.

Herndl, Carl G. 1996. Tactics and the quotidian: Resistance and professional discourse. *Journal of Advanced Composition* 16 (3): 455–470.

Horner, Bruce. In press. Critical ethnography, ethics, and work: Re-articulating labor. In *Protean Ground: Critical Literacy and the Postmodern Turn*, eds. Stephen Brown and Sydney Dobrin. Albany: State University of New York Press.

Hull, Glynda. 1997. Hearing other voices: A critical assessment of popular views on literacy and work. In *Changing work, changing workers: Critical perspectives on language, literacy and skills*, ed. Glynda Hull, 3–39. Albany: State University of New York Press.

———. 1999. What's in a label? Complicating notions of the skills-poor worker. *Written Communication* 16(4): 379–411.

Hutchins, Edwin. 1993. Learning to navigate. In *Understanding practice: Perspectives on activity and context*, eds. Seth Chaiklin and Jean Lave, 35–63. Cambridge: Cambridge University Press.

————. 1996. *Cognition in the wild*. Cambridge: Massachusetts Institute of Technology Press.

Katz, Susan M. 1998. A newcomer gains power: An analysis of the role of rhetorical expertise. *Journal of Business Communication* 35 (4): 419–442.

————. 1999. *The dynamics of writing review: Opportunities for growth and change in the workplace*. Stamford, CT: Ablex.

Latour, Bruno. 1987. *Science in action: How to follow scientists and engineers through society*. Cambridge: Harvard University Press.

————. 1990. Drawing things together. In *Representation in scientific practice*, eds. Michael Lynch and Steve Woolgar, 19–68. Cambridge: Massachusetts Institute of Technology Press.

————. 1996. *Aramis or the love of technology*. Translated by Catherine Porter. Cambridge: Harvard University Press.

————. 1999. *Pandora's hope: Essays on the reality of science studies*. Cambridge, MA: Harvard University Press.

Lave, Jean. 1995. The practice of learning. In *Understanding practice: Perspectives on activity and context*, eds. Seth Chaiklin and Jean Lave, 3–32. Cambridge: Cambridge University Press.

Lave, Jean, and Etienne Wenger. 1991. *Situated learning: Legitimate peripheral participation*. Cambridge: Cambridge University Press.

Law, John. 1994. *Organizing modernity*. Cambridge, MA: Blackwell.

Leana, Carrie R., and Harry J. Van Buren, III. 1999. Organizational social capital and employment practices. *Academy of Management Review* 24 (3): 538–555.

Lewis, Laurie K. 2000. Communicating change: Four cases of quality programs. *Journal of Business Communication* 37 (2): 128–155.

Longo, Bernadette. 2000. *Spurious coin: A history of science, management, and technical writing*. Albany: State University of New York Press.

MacKinnon, Jamie. 1993. Becoming a rhetor: Developing writing ability in a mature, writing-intensive organization. In *Writing in the workplace: New research perspectives*, ed. Rachel Spilka, 41–55. Carbondale: Southern Illinois University Press.

Mangrum, Faye Gothard, Michael S. Fairley, and D. Lawrence Wieder. 2001. Informal problem solving in the technology-mediated work place. *Journal of Business Communication* 38 (3): 315–336.

Marvin, Carolyn. 1988. *When old technologies were new: Thinking about electrical communication in the late nineteenth century*. New York: Oxford University Press.

Massey, Joseph Eric. 2001. Managing organizational legitimacy: Communication strategies for organizations in crisis. *Journal of Business Communication* 38 (2): 153–182.

Medway, Peter. 1996. Virtual and material buildings: Construction and constructivism in architecture and writing. *Written Communication* 13 (4): 473–514.

Middleton, David. 1996. Talking work: Argument, common knowledge, and improvisation in teamwork. In *Cognition and communication at work*, eds. Yrjo Engstrom and David Middleton, 233–256. Cambridge: Cambridge University Press.

Miettinen, Reijo. 1998. Object construction and networks in research work: The case of research on cellulose-degrading enzymes. *Social Studies of Science* 28 (3): 423–463.

Miles, Matthew B., and A. Michael Huberman. 1994. *Qualitative data analysis*. Thousand Oaks, CA: Sage.

Miller, Carolyn R. 1984. Genre as social action. *Quarterly Journal of Speech* 70 (2): 151–167.

Muckelbauer, John. 2000. On reading differently: Through Foucault's resistance. *College English* 63 (1): 71–94.

Mukerji, Chandra. 1996. The collective construction of scientific genius. In *Cognition and communication at work*, eds. Yrjo Engstrom and David Middleton, 257–278. Cambridge: Cambridge University Press.

Nevis, Edward R., Anthony J. DiBella, and Janet M. Gould. 1995. Understanding organizations as learning systems. *Sloan Management Review* 36 (2): 73–85.

Newkirk, Thomas. 1996. Seduction and betrayal in qualitative research. In *Ethics and representation in qualitative studies of literacy*, eds. Peter Mortensen and Gesa E. Kirsch, 3–16. Urbana, IL: National Council of Teachers of English.

Paradis, James, David Dobrin, and Richard Miller. 1985. Writing at Exxon ITD: Notes on the writing environment of an R&D organization. In *Writing in nonacademic settings*, eds. Lee Odell and Dixie Goswami, 281–307. New York: Guilford.

Pare, Anthony. 1993. Discourse regulation and the production of knowledge. In *Writing in the workplace: New research perspec-*

tives, ed. Rachel Spilka, 111–123. Carbondale: Southern Illinois University Press.

———. 2000. Writing as a way into social work: Genre sets, genre systems, and distributed cognition. In *Transitions: Writing in academic and workplace settings*, eds. Patrick Dias and Anthony Pare, 145–166. Cresskill, NJ: Hampton Press.

Putnam, Robert. 1993. The prosperous community: Social capital and public life. *American Prospect* 4 (13): 35–42.

Sauer, Beverly. 1998. Embodied knowledge: The textual representation of embodied sensory information in a dynamic and uncertain material environment. *Written Communication* 15 (2): 131–169.

Schryer, Catherine F. 1993. Records as genre. *Written Communication* 10 (2): 200–234.

Shapin, Steven. 1989. The invisible technician. *American Scientist* 77 (6): 554–563.

Smart, Barry. 1985. *Michel Foucault*. New York: Routledge.

Smart, Graham. 1993. Genre as community invention: A central bank's response to its executives' expectations as readers. In *Writing in the workplace: New research perspectives*, ed. Rachel Spilka, 124–140. Carbondale: Southern Illinois University Press.

Strauss, Anselm, and Juliet Corbin. 1990. *Basics of qualitative research: Grounded theory procedures and techniques*. Newbury Park, CA: Sage.

Sullivan, Francis J. 1997. Dysfunctional workers, functional texts: The transformation of work in institutional procedure manuals. *Written Communication* 14 (3): 313–359.

Swales, John. 1990. *Genre analysis: English in academic and research settings*. Cambridge: Cambridge University Press.

Thompson, John B. 1991. Editor's introduction to *Language and symbolic power* by Pierre Bourdieu. Cambridge: Harvard University Press.

Tillery, Denise. 2001. Power, language, and professional choices: A hermeneutic approach to teaching technical communication. *Technical Communication Quarterly* 10(1): 97–116.

Unseem, M., and J. Karabel. 1986. Pathways to top corporate management. *American Sociological Review* 44 (1): 184–200.

Vincenti, Walter G. 1990. *What engineers know and how they know it: Analytical studies from aeronautical history*. Baltimore: Johns Hopkins University Press.

Vygotsky, Lev S. 1978. *Mind in society: The development of higher psychological processes*. Edited by Michael Cole, Vera John-Steiner, Sylvia Scribner, and Ellen Souberman. Cambridge: Harvard University Press.

Wehner, Pat. 2001. Opinion: Ivory arches and golden towers: Why we're all consumer researchers now. *College English* 63 (6): 759–768.

Wenger, Etienne. 1998. *Communities of practice: Learning, meaning, and identity*. Cambridge: Cambridge University Press.

Whalley, Peter, and Stephen R. Barley. 1997. Technical work in the division of labor: Stalking the wily anomaly. In *Between craft and science: Technical work in U.S. settings*, eds. Stephen R. Barley and Julian E. Orr, 23–52. Ithaca: Cornell University Press.

Winsor, Dorothy A. 1989. An engineer's writing and the corporate construction of knowledge. *Written Communication* 6 (3): 270–285.

———. 1990a. Engineering writing/writing engineering. *College Composition and Communication* 41 (1): 58–70.

———. 1990b. How companies affect the writing of young engineers: Two case studies. *IEEE Transactions on Professional Communication* 33 (3): 124–129.

———. 1992. What counts as writing? An argument from engineers' practice. *Journal of Advanced Composition* 12 (2): 337–347.

———. 1994. Invention and writing in technical work: Representing the object. *Written Communication* 11 (2): 227–250.

———. 1996. *Writing like an engineer: A rhetorical education*. Mahwah, NJ: Erlbaum.

———. 1999. Genre and activity systems: The role of documentation in maintaining and changing engineering activity systems. *Written Communication* 16 (2): 200–24.

Yates, JoAnne. 1989. *Control through communication: The rise of system in American management*. Baltimore: Johns Hopkins University Press.

Yates, JoAnne, and Wanda Orlikowski. 2002. Genre systems: Structuring interaction through communicative norms. *Journal of Business Communication* 39 (1): 13–35.

Zuboff, Shoshana. 1988. *In the age of the smart machine: The future of work and power*. New York: Basic.

Index

171